T0224042

Reflections on Human Inquiry

Nirmalangshu Mukherji

Reflections on Human Inquiry

Science, Philosophy, and Common Life

Springer

Nirmalangshu Mukherji
University of Delhi
Delhi
India

ISBN 978-981-13-5378-9   ISBN 978-981-10-5364-1   (eBook)
DOI 10.1007/978-981-10-5364-1

Printed on acid-free paper

This Springer imprint is published by Springer Nature
The registered company is Springer Nature Singapore Pte Ltd.
The registered company address is: 152 Beach Road, #21-01/04 Gateway East, Singapore 189721, Singapore

Criticism has torn up the imaginary flowers from the chain not so that man shall wear the unadorned, bleak chain but so that he will shake off the chain and pluck the living flower.

Karl Marx ('Contribution to the Critique of Hegel's Philosophy of Law', in K. Marx and F. Engels, *On Religion*, Moscow: Progress Publishers, pp. 38–52)

Acknowledgements

Philosophical writing broadly divides into two modes: literary and scientific. These philosophical essays were written over the last two decades to explore the literary mode of our common lives as a counterpoint to the scientific mode that otherwise occupies my academic life. They remained scattered, unrelated to each other, in obscure journals and anthologies for many years; in some cases, they remained unpublished. I am pleased with the prospect that they may now be read together under the noted general perspective. I have reworked all of them to enforce some form of unity, however loose, for inclusion in this volume. Some of them are freshly written for the volume and are published here for the first time. Most of the essays which were published before are now radically altered from their printed versions; only a few retain significant resemblance with their past.

As these were presented to a variety of audiences over the last decade, I have benefitted from exchanges with too large a number of people to mention here. But I must mention my close friend, the elegant philosopher Bijoy Boruah, who was not only present on many occasions in which these essays were presented, he has discussed almost all of them with me over the years as the idea of this volume slowly took shape. I particularly remember a long drive on the highways to the Himalayas, engrossed in the problem of whether our modes of knowledge cloud our conception of reality. Unsurprisingly, we lost our way and ended up in a different province. I must also thank Samiksha Goyal for helping me with the preparation of the manuscript.

As noted, most of the essays collected in this volume are thoroughly revised versions of articles published earlier. The original publications, if any, are noted at the beginning of the essay. I am thankful to the editors for granting me permission to include these essays in this volume.

Contents

About the Author

Nirmalangshu Mukherji is a former Professor of Philosophy at the University of Delhi. He was National Visiting Professor for the Indian Council of Philosophical Research (2015–16). His primary academic interest is the study of language and mind. His publications in this broad area include *The Cartesian Mind: Reflections on Language and Music* (2000), *The Primacy of Grammar* (2010), *and Language, Music and Cartesian Mind: Reflections on Grammar of Cognition* (In Press). He also co-edited Noam Chomsky's *The Architecture of Language* (2000). Prof. Mukherji is also professionally interested in the nature of human inquiry, including the character of philosophical practice. Some of his work in this area is collected in the present book. He is actively engaged with issues of peace, justice and human rights. Apart from numerous articles, he has published two books in this area: *December 13: Terror over Democracy* (2005) and *Maoists in India: Tribals under Siege* (2012).

Chapter 1
Introduction

1.1 Reflective Pluralism

Human beings are endowed with cognitive agency. Our grasp of the world, and of ourselves, is not merely a reflexive response to external stimuli, but also a reflective product of human inquiry, often structured in imagination. What are the forms of inquiry available to humans to lead a significant life? How are these forms related to each other? The 11 exploratory chapters of varying length collected in this volume examine forms and limits of human inquiry from a variety of directions.

Most of these directions emanate from classical philosophical investigations on human knowledge. Since the nature of human inquiry is the general theme, it is unsurprising that the chapters cover a wide range of familiar philosophical topics: the nature of reality, scientific realism; concepts of truth, knowledge, belief, consciousness; character of mind, language, grammar, meaning; literature and philosophy; nature of music, religious discourse; knowledge and human destiny, and others. Although I have called them 'chapters', it is not unreasonable to view the volume as a collection of essays.

These pieces were written in a discontinuous fashion over a number of years for very different occasions and audiences, and at varying, often conflicting, reflective moments. Strictly speaking, their spatial assembly here does not really amount to a sustained fully articulated monograph; significant silences insulate the individual write-ups from each other. Given the range and complexity of the listed topics, I do not think there could be a single substantive monograph that covers them all. In any case, I am not concerned here either with history of philosophy or with philosophical anthropology, even though I end up doing these things on occasion to set the scene. My intention is not to report on the current state of these topics. They are discussed because they necessarily infiltrate the mind when you think about the idea of being human.

Yet this is not just a compilation of assorted papers to mark the end of a career. If a metaphor is needed to cover the collation, one could say, in celebrated terms, that

© Springer Nature Singapore Pte Ltd 2017
N. Mukherji, *Reflections on Human Inquiry*,
DOI 10.1007/978-981-10-5364-1_1

they form a 'family' as their resemblances 'criss-cross and overlap'. I think it is better to view the pieces as forming a group of proximate islands in the same stretch of the sea; the image of an island seems appropriate because each chapter stands on its own without directly depending on the others. However, I have used the method of cross-reference frequently to aid the memory, sharpen a point, or to construct a bridge. I will try to describe the composite picture shortly.

Individually, too, the pieces are more like free-flowing essays than formally structured papers meant for disciplinary journals. I am aware that centuries of the most extensive reflection and scholarship across many fields of inquiry have nourished each of the topics listed above. Especially in the last century almost all of these topics have attained formidable technical character. Apart from developing theoretical vocabulary of their own, philosophers have explored these issues with insights from mathematical logic, theoretical linguistics, computer science, cognitive psychology, neuroscience and theoretical physics. As a result, it is now expected that these otherwise large and elusive issues are discussed in terms of the latest technical proposal; fair enough, that is how academic papers are written.

The pieces assembled here generally do not follow that trajectory. Although they do cover familiar philosophical topics like knowledge, truth, realism, belief, meaning, interpretation and the like, which are often discussed in professional platforms, these topics carry much value beyond the closely guarded canons of the academia. After all, as the legend goes, many of these topics started their career on ancient streets or under banyan trees; arguably, unlike other branches of inquiry, they retain the memory of those plebeian assemblies. These chapters attempt to convey a sense of relaxed conversation in a disarming voice to reach audiences outside professional meetings of philosophers. As a result, they sometimes ignore, or even disobey, the formal tone and attire of academic discourse.

However, these are not 'popular' pieces by any means. After a life in professional philosophy, often guided by inputs from the adjacent sciences, it is by now intellectually impossible to entirely avoid the formal tone and at least some of the demanding literature that informs it. In that slightly uncertain sense, these are reflective efforts that are seeking a zone of comfort somewhere between technical journals and literary supplements, but never aiming for a talk-show. As a result, in many cases, they start out with the usual preparations of the professional philosopher, but they seldom stay on course to the end; in a variety of ways, the discussion moves away from familiar abstract channels to more direct arenas of common life. It is not for me to judge whether the effort had been successful, but I hope they do convey some sense of honesty of purpose because, in most cases, the discourse was not deliberately designed.

1.1.1   From a Sceptical Point of View

So, what explains the diffused character of these chapters? I think the answer lies in the way in which my own intellectual interests unfolded. Having made a decision to shift, early in my career, from the beautiful abstractions of mathematical physics to the more existential concerns of philosophy, I settled down to a range of exciting new developments in analytic philosophy in the post-Wittgensteinian era. The work of fine philosophers like John Austin, Peter Strawson, Willard Quine, Hilary Putnam, Donald Davidson, Michael Dummett, and other stalwarts of late twentieth-century analytic philosophy, promised a healthy mix of rigorous, often formal, inquiry with what Hilary Putnam called 'the whole hurly-burly of human actions' (cited in Nussbaum 2016). Philosophers such as Strawson (1992) and others have often suggested that philosophy attempts to produce a systematic account of the general conceptual apparatus of which our daily practices display a tacit and unconscious mastery.

But the subtle, abstract and yet unifying framework of physics lingered in the mind. This led to a variety of dissatisfaction with analytic philosophy, especially in the study of language. We need to step back a little to see why. In the first half of the twentieth century, great philosophers like Gottlob Frege, Bertrand Russell, Ludwig Wittgenstein, Rudolf Carnap, Alfred Ayer and others took what Rorty (1967) called the *linguistic turn*. Tracing it to the philosophy of Immanuel Kant, Coffa (1991) called this mode of doing philosophy the *semantic tradition*. Within this broad tradition, each of the authors cited in the preceding paragraph—Austin, Quine, etc.—belonged primarily to the broad discipline of philosophy of language. The study of language thus formed a central part of the analytic effort. As with most students of analytic philosophy in those days, I was attracted to the study of language both for the intricate formal character of human language, and its ubiquitous role in human life.

Linguistic philosophy promised a rigorous, scientific approach of its own on classical philosophical topics such as realism, knowledge, belief, even consciousness. For example, Quine (1953) argued that for something to exist it has to be the value of a bound variable in a true theory; Wittgenstein (1953) suggested that to understand consciousness is to understand the meaning of the first-person sentence *I am in pain*; Russell (1919) held that beliefs such as <Ramanuj is wise> are propositional attitudes. I will have much more to say on these things in the chapters that follow.

Since linguistic philosophy proposed to examine classical issues by viewing them as 'semantic' problems—that is, in terms of the structure and function of language—it is reasonable to expect that this philosophy will also furnish a formally satisfying account of language itself from which the solution to philosophical problems maybe rigorously derived. However, linguistic philosophy lacked a genuinely theoretical understanding of the immense richness of human language. This is what a mind initially trained in physics sorely missed. This philosophy did make formal proposals occasionally, such as Russell's famous theory of descriptions (Russell 1905), to address philosophical problems. But the formal tools were borrowed from the discipline of symbolic logic which is not only a poor substitute for human language; its character is parasitic on human language.

In any case, even with the tools of formal logic, human language resisted any grand formal theory for addressing philosophical problems, as Strawson (1950) pointed out in his stringent criticism of Russell's theory of descriptions: ordinary language, Strawson declared, has no logic. 'Ordinary language' philosophers thus focused on detailed, taxonomic properties of language in the style of a botanist, as Austin (1962) suggested, rather than that of a physicist. The study of language fostered what Strawson (1971) called a *Homeric struggle* between 'formal-semantic' and 'communication-intention' theorists of language. My impression is that the scene in analytic philosophy hasn't improved since even if no one openly makes claims for either 'ideal language' or 'ordinary language'. At that stage, it was too early for me to admire the value of this uncertainty in philosophical inquiry.

While analytic philosophy was going through this apparent absence of direction, interesting developments took place elsewhere. I expressed my disenchantment with the state of linguistic philosophy in my doctoral thesis, and turned to linguistics and cognitive science to see if there was a 'physics' of human language and mind. Two related developments promised what I was looking for: exciting proposals in theoretical linguistics by Noam Chomsky, and the formulation of a computational theory of mind by Alan Turing. Both strands of research, and much else besides, had become established academic pursuits by the time I completed my doctoral thesis. As I continued with my exploration of the new science of the mind, certain interesting ideas and results did appear on the table in due course which I put together in some papers and monographs culminating in *The Primacy of Grammar* (Mukherji 2010). That form of work continues elsewhere.

However, throughout my engagement with the new science of the mind, I was beginning to realize that the ideas that interested me there covered very restricted and abstract domains of human cognition such that the intellectual salience of much of the rest of the new science could be questioned. For example, the formal resources of linguistic theory no doubt explained some intriguing facts about how sound is connected to what may be called the *internal significance* of a structure, called *Logical Form* (LF) in the technical literature. However, it is also clear that the theory does not have either the resources or the desire to explain what may ordinarily be viewed as the meaning of a sentence.

Where does the rest of the meaning come from to enrich human cognition? Needless to say, ever more sophisticated investigations on the nature of human language are under way to expand the scope of linguistic theory and to address the doubts just raised (Hinzen and Sheehan 2013). Yet, as argued in *The Primacy of Grammar*, it is not evident if any significant notion of theory applies beyond grammatical investigations. As far as genuinely scientific studies on language go, there is grammatical theory stuck at LF, and there is philately.

Given the predominance of language in human cognitive architecture, the restriction just sketched seems to be the case for much cognitive investigation as well where language is intimately involved: in the study of concepts and reasoning, for example. For the rest of the cognitive studies detached from language, the scene seems to be worse since there is no sign of 'physics' at all; it is mostly just fancy organization of behavioural data. Thus there is much room for wide-ranging scepticism about the scope of the cognitive sciences. The Homeric struggle seemed to extend far beyond language; it threatened to cover the architecture of human cognition itself. It seemed that not only that the botanist plays a crucial role in human inquiry; there are also areas of deep human concern to which even the botanist does not have access. Yet humans tread those areas with impressive cognitive confidence as they lead their common lives.

There seems to be three options in hand with respect to how we respond to the sceptic. First, one could keep digging at the vast phenomenon of human cognition with whatever scientific tool is in hand; this is what cognitive scientists and philosophers are doing in any case. Some of my own continuing work falls under this option, as noted; we may ask, for instance, if language and music share the same grammatical structure. Second, one could embrace wholesale scepticism about science, refuse to make any formal-theoretical move, and turn philosophical problems into 'literary' activities: call it *post-structuralism*. Third, one uses scepticism as a strategy to progressively expand the notion of human inquiry; in other words, by showing the limitations of one form of inquiry one draws attention to the significance of some other forms. In effect, we may view alternative forms of inquiry as reinforcing—rather than negating—each other: call it, if you like, *reflective pluralism*.

I don't think that the chapters that follow mark any definite choice between these broad options, for reasons—including moral and political ones—that emerge as we proceed. Basically, the inclination is to leave things as they are. However, it will not be implausible to detect sympathy for the first and the third options, and an attempt to come to terms with their 'incommensurability'. There is also a tendency to ignore the second option largely because holding it along with the other two options precipitates flat inconsistency; hence, I have ignored the vast literature—Roland Barthes, Michel Foucault, Richard Rorty, Jacques Derrida and others—that propagates the second option. Moreover, the second option grants salience to just one form of inquiry, namely, the literary one; after spending a life in analytic philosophy and in admiration of physics, one develops a visceral discomfort with any proclamation that fails to uphold their value. But the association with the formal does not prevent me to shift to the literary mode whenever needed.

In any case, I lack the enthusiasm to *argue* these choices here because I have very little interest in metatheory. I rather prefer Wittgenstein's idea of simply *describing* the modes of human inquiry—'forms of life,' as he would say—as they shore up when we look for them within the vicinity of our own agency. In any case, notwithstanding the option one recommends, there is the need to furnish something of a perspective for the phenomenon that humans have reflective resources to lead cognitively meaningful lives. What are those resources? Is there an account of human cognitive agency as a whole?

These chapters started emerging one after another as a variety of very specific questions about the form and limits of human inquiry began to form in mind. For example, at one point in human history it was thought that modern science, especially theoretical physics, is the paradigm of human inquiry. Where does this form of inquiry significantly apply? Are there limits on its claims of truth and objectivity? How much of the vast canvas of human experience does it cover? Where do other forms of inquiry, such as philosophy, literature, religion, and the arts, attain their salience?

With the emergence of scientific study of the human mind itself, these critical questions have taken a more intriguing form, as noted. Can human inquiry investigate its own nature? Can the scientific theory of language explain the richness of human expression? Can a science of the mind account for human experience? These probing questions on the scientific enterprise are usually addressed from the outside, as it were, by humanists, philosophers of science, sociologists of knowledge and critical theorists. In these chapters, they are examined from the inside by a philosopher whose primary academic work concerns the study of the human linguistic mind. In that sense, the sceptical inquiry turns on itself.

1.1.2 The Chapters

Each chapter in this volume is accompanied by a substantial abstract that lays out the theme of the chapter. What I plan to do now is to give some idea of the family of concerns that link these chapters in a variety of ways. As noted, the starting point of this exercise is the idea of science. When we face the entirety of human inquiry in its kaleidoscopic state, we need some categories to describe the spectacle. The idea of science seems to offer that handle. Modern science represented a very classical conception of human knowledge as an objective quest for the real properties of the world. With its grand mathematical architectonic, physics was able to develop tools of investigation that unearthed deeply hidden features of the universe. But its highly esoteric form of discourse and extremely theory-internal conception of the world makes physics unavailable to the general cognitive agent, including the physicist outside his specialist forum. With the advent of modern science then it looks as if humans engage in two basic forms of inquiry: let us call them *scientific* and *cultural*, respectively. As we will see in the chapters that follow, the labels themselves are of less value than details about the underlying forms.

In the scientific mode, human inquiry claims knowledge of reality: the knowledge constitutes the truth-claims of science, and the reality constitutes the joints of nature so postulated. The discourse is assumed to be absolute and objective. The truth-claim no doubt is a human action, but the truth—such as the Earth is round—is independent of any agent, community, tradition, textual and social context; in other words, truth lays bare the world as it is. It is commonly believed that the scientific conception of the world is *objective* in the sense that it does not have a (preferred) point of view; Nagel (1986) called it the *view from nowhere*.

In contrast, much of our lives includes a *subjective* point of view, the point of view of the human agent; these may be thought of as *views from somewhere*. As Nagel (1986) and Davidson (1991) pointed out, the two views need to be reconciled in order for us to lead a meaningful life including social and political lives. Nagel then goes on to show how the reconciliation is to be achieved to address a range of classical philosophical problems, such as the mind–body problem. Speaking roughly, the distinction between *view from nowhere* and *view from somewhere* is one way of formulating the distinction between the scientific and the cultural.

My interests are markedly different from the suggested distinction. I think there is another distinction between the scientific and the cultural which is related to, but not sufficiently captured by, the subjective–objective distinction. As noted, both the subjective and the objective perspectives are needed to reach human thought and action (Davidson explicitly adds the inter-subjective perspective to the other two); human thought is the result of a *reconciliation* of these things in any case. I think a scientific-cultural distinction arises even after such reconciliation is reached. Chapters 2 and 3 in this volume discuss the issue.

The starting point is the conception of knowledge. In Chap. 2 ('Human Reality'), I show how the concepts of knowledge, truth and reality are intimately related; if a conception of mind-independent reality is unavailable, so are the concepts of knowledge and truth. The problem is that human knowledge and, therefore, the conception of reality are necessarily products of how humans are designed; if humans were designed, say, as bats, the conception of the world would have been very different. So, if the notion of objectivity is understood in terms of a mind-independent reality, then that notion appears to be problematic, if not downright incoherent. There is much room for scepticism then regarding realist claims. *Within* the design, though, it is striking that the human mind can sometimes detect formal/mathematical regularity in the external world. The phenomenon is poorly understood but its shining existence cannot be denied. Perhaps it is possible to recover some version of the notions of knowledge, truth and reality around this phenomenon. I discuss the possibility with more constructive details in Chap. 3.

However, the formal mode of inquiry is rarely available in the vast stretch of human cognitive life. This suggests a broad distinction between forms of inquiry regarding the presence and absence of the formal mode, which amounts roughly to the distinction between the scientific and the cultural. It could be that the world and the knowledge of it are reached in very different reflective terms between the two forms of inquiry. In that sense, the world lost in our analytic pursuit may be regained in our poetic form of inquiry in which the world is grasped by immersing ourselves in it. The elusive world, which we are unable to discover except in rare cases by looking at it from the outside, is cheerfully embraced as a lived world from the inside.

Chapter 3 ('Science and the Mind') focuses on the historical fact that the scientific mode is a great human achievement, but it works in very restricted domains of simple systems. That's the price we pay for our penchant for objectivity. Genuine scientific understanding is reached primarily through the formal mode—the Galilean style—which is available only for very simple systems. The chapter points out that the arts also sometimes search for formal/minimalist conception of aspects of the world, but the method of search is distinct, resulting in a vastly different form of inquiry. It is reasonable to expect then that a genuine science of the mind is also likely to be restricted only to those aspects of the mind where the formal mode is available. Human language is perhaps the most promising example of such an aspect of the mind. However, there are serious limits to the inquiry even there, as the next two chapters suggest.

Chapter 4 ('Theories and Shifting Domains') examines the sense in which scientific theories in the formal mode identify a stretch of the world. The contemporary discipline of linguistic theory is an interesting example to study in this context because of its recentness; we are able to study its entire history in a stretch to see whether the reality of human language has come into sharper focus as the theory progressed. After a brief exposition of the basic joints of the theory, it turns out that, even within its short history, the object of the theory has become increasingly theory laden for Chomsky (1991) to remark that perhaps there is no such thing as language.

Chapter 5 ('The Sceptic and the Cognitivist') adds another dimension to the scepticism just raised. This chapter joins issue with recent claims from the cognitive sciences that the ancient discipline of philosophy is beginning to lose its relevance for understanding human cognition. We focus again on the new discipline of linguistic theory, which is perhaps the most promising programme in the cognitive sciences. As the work of philosophers of language mentioned earlier highlighted, the basic classical interest in the study of language has been that humans have the astonishing ability to talk about the world: the *semantic* ability. As hinted earlier, the theoretical resources of linguistic theory seems to fall far short of the philosophical interest.

Having secured something like a zone of autonomy for the philosophical form of inquiry in Chap. 5, Chap. 6 ('From Things to Needs') attempts to develop the idea of autonomy by focusing on the general form of classical Indian philosophy. It may be justly complained that, unlike Western philosophy, this philosophy has lost its relevance because it never interacted with the vast edifice of European science. This conclusion will follow only under the assumption that scientific knowledge overrules or replaces philosophical inquiry. A quick look at the origin and form of Indian philosophy suggests that its goals might not have been to discover properties of the world at all. A salient goal for philosophical inquiry, distinct from the sciences, could be to formulate conditions of human reflective *needs* for cognitive agents to lead rational lives. The study of needs seems to be fundamental to philosophical inquiry since its presence can be located even in classical Western philosophy when it is shorn of its 'scientific' goals. Interestingly, the study of the mind—the contentious domain under consideration—offers some promising evidence on this issue. In this light, each of the concepts of consciousness, knowledge and belief may be understood very differently from their alleged 'mentalistic' features discussed in the received literature.

Chapters 7–9 ('Yearning for Consciousness', 'Ascription of Knowledge', 'Beliefs and Believers') cover the alternative perspective. The chapters exploit the general distinction between *description* and *ascription*. While the goal of descriptions is to examine properties of objects, ascriptions suggest devices of personal evaluation. Each chapter thus consists of two distinct sections. In the first section, we show that the current state of philosophical inquiry on these concepts is at best uncertain; there appear to be fundamental conceptual darkness around them. However, each concept turns out to be salient when we think of them as recommending different evaluative attitudes towards persons and communities to enable us to get a grip on our interpersonal lives.

The idea of placing much of philosophical inquiry into the cultural mode raises the issue of whether the notion of the cultural, as distinct from the scientific, is a coherent unified category. One way of examining the issue is to locate some invariant notion of interpretation governing each of the putative cultural objects. A somewhat detailed 'anthropological' study pursued in Chap. 10 ('Varieties of Interpretation') across rituals, poetry, painting and music suggests that even the notion of interpretation radically varies as the objects vary. So, for example, we cannot say without equivocation that cultural objects have a distinctive aspect in that they admit of both singular and plural interpretations.

The perspectives that govern interpretations come in a variety of forms: plurality of traditions, bounds of space and time, eras and epochs, textuality and interpretations, multiplicity of languages, gestalt properties, and simply differences of irreconcilable opinion, often assuming the form of class war. None of these are seen in science, say, in theoretical physics. No doubt, there are scientific disputes, but that is a different matter altogether. Beyond this general observation of open-ended plurality, human inquiry is too diffused an undertaking to lend itself to definite categories.

Yet we can locate on examination that there are tangible distinctions between forms of inquiry, even if they blend into one another to mask their identity. For example, we could make some sense of the distinction between the scientific and the philosophical modes as above even if philosophical inquiry sometimes takes a scientific form up to a point. Similarly, there is a perceived sense of affinity between philosophy and literature as an impressive body of 'converging' literature testifies. Focusing on the non-converging literature, Chap. 11 ('Literature and Common Life') takes up one of the leading issues raised in this volume: where does common life get its enrichment from in the general absence of scientific reflection? The answer projected in this chapter appeals to the notion of a *text*. An author's view from somewhere enshrined in a text—Platonic or Shakespearean—enables the cognitive agent to expand her horizons and transcend her locality.

Given the variety, richness, and autonomy of forms of human inquiry, it is difficult—perhaps even morally questionable—to prioritize a specific form of knowledge. In any case, as we saw, even what is taken to be the pinnacle of human inquiry, namely, formal science, has only limited role in human life. In this essentially pluralistic conception of human knowledge, Chap. 12 ('Education for the Species') raises the issue of the value of this edifice of human knowledge. Sketching the grim scenario for the survival of the human species, it is argued that much of the damage can be traced to the adoption of highly prioritized knowledge-systems ensuing from elite high cultures. In contrast, the marginalized knowledge-systems of the indigenous people across the world offer a salient perspective for saving the planet. The salience of indigenous knowledge entails a large-scale rejection of elite knowledge-systems. If scepticism is viewed as a state of mind that rejects dominating knowledge-systems, humans need to adopt probably the most extreme form of scepticism, if the species is to survive.

References

Austin, J. 1962. In *How To Do Things With Words*, J. Urmson (Ed.). Oxford: Oxford University Press.

Chomsky, N. 1991. Linguistics and cognitive science: problems and mysteries. In *The Chomskyan Turn*, A. Kasher (Ed.), 26–53. Oxford: Basil Blackwell.

Coffa, A. 1991. *The Semantic Tradition: From Kant to Carnap To the Vienna Station*. Cambridge: Cambridge University Press.

Davidson, D. 1991. Three varies of knowledge. In *Royal Institute of Philosophy Supplement*, A. Phillips Griffiths (Ed.), 153–166. New York: Cambridge University Press.

Hinzen, W., and M. Sheehan. 2013. *The Philosophy of Universal Grammar*. Oxford: Oxford University Press.

Mukherji, N. 2010. *The Primacy of Grammar*. Boston: MIT Press.

Nagel, T. 1986. *The View from Nowhere*. New York: Oxford University Press.

Nussbaum, M. 2016. Hilary Putnam (1926–2016). *Huffington Post*, 14 March.

Quine, W. 1953. On what there is. In *From a Logical Point of View*. Cambridge: Harvard University Press.

Rorty, R. 1967. *The Linguistic Turn*. Chicago: University of Chicago Press.

Russell, B. 1905. On denoting. *Mind* 14: 479–493.

Russell, B. 1919. The philosophy of logical atomism. *Monist*. In *Logic and Knowledge* (Reprinted, 1956), R. Marsh (Ed.), 177–281. London: George Allen and Unwin.

Strawson, P. 1950. On referring. *Mind*, July. Reprinted in Strawson (1971).

Strawson, P. 1971. *Logico-linguistic Papers*. London: Methuen.

Strawson, P. 1992. *Analysis and Metaphysics*. London: Oxford University Press.

Wittgenstein, L. 1953. *Philosophical Investigations*. Translated by G. Anscombe. Oxford: Blackwell Publishers.

Chapter 2
Human Reality

Where indeterminacy of translation applies, there is no real question of right choice; there is no fact of the matter even to within the acknowledged under-determination of a theory of nature.

Willard Quine.

2.1 Is There a World?

Realism is the view that objects and events in the world exist independently of human conception of them: the world is mind-independent. Looking at the book in front of me, I am prone to view the book as existing on its own even when I am the author of the book. That is, even when I know that the book could not have come to exist without my mental and physical effort, the existence of the book itself is not dependent on my mind. The book will continue to exist when I am no longer looking at it and thinking about it.

Needless to say, our sense of realism is more robust for unambiguously external objects such as rivers, trees and mountains. Much of human enterprise is geared to come to terms with this external world, including the social world, so that we are able to lead a meaningful—and hopefully happy—life in it. We engage with this world for every breath we take, every move we make, every step we take. The effort would have been far less demanding and interesting if the world happened to be just a manifestation of our minds. We live in the world even when we crave for the utopia. This much seems to be evident even if set aside the Cartesian logical problem that my own existence becomes problematic if the world isn't there.

On closer reflection though, the opposite anti-realist idea—that the world is fundamentally our own construction—seems to have considerable force. Classical philosophers, especially in the rationalist tradition, pointed out that humans are not just passive receivers of external stimuli. The human mind actively contributes from its own inner resources to organize and interpret sensory information. The rationalist philosopher Ralph Cudworth (1731) called these resources 'cognoscitive powers' which enable the mind to raise 'intelligible ideas and conceptions of things

© Springer Nature Singapore Pte Ltd 2017
N. Mukherji, *Reflections on Human Inquiry*,
DOI 10.1007/978-981-10-5364-1_2

from within itself'. The 'intelligible forms by which things are understood or known', Cudworth held, 'are not stamps or impressions passively printed upon the soul from without, but ideas vitally protended or actively exerted from within itself.' For another rationalist philosopher René Descartes, the human ability to form conceptions of things from chaotic and often-impoverished experience is akin to the formation of 'a statue of Mercury contained in a rough block of wood'. Thinking of human knowledge on the analogy of the statue of Mercury, it becomes very unclear if any significant notion of mind-independent reality can be attached to the statue. In an uncomfortably strong sense, then, the conception of reality can only be a conception of human reality.

To dispel possible misinterpretations, I distinguish between four senses of *human reality* to suggest that only the last one is under discussion here. First, by *human reality* people sometimes mean that reality has human-like properties much like humanoid faces in clouds. Various animistic conceptions of reality, such as panpsychism, often invoke anthropomorphic properties of reality itself. Second, in anthropological inquiry, the notion of a human reality sometimes indicates those aspects of the world—buildings, bridges, ships, computers, books and artworks— that are clearly products of human effort. Third, by *human reality* we could mean the conditions that humans face as socio-biological creatures: oppression, political history, and inevitability of death.

None of these three senses is at issue in this chapter. By *human reality*, I simply mean the reality that humans grasp by dint of their unique cognoscitive powers. Despite its metaphysical import, the notion of human reality is epistemologically linked to the human agent: the kind of reality humans know of. The notion is more fundamental than the other notions of human reality since the former subsumes the latter; for example, the reality humans grasp by dint of their epistemic powers could well have animistic features.

2.2   Truth and Knowledge

However, it is not often realized in intellectual circles outside analytic philosophy that two rather standard ideas in philosophy lead to the conclusion that the object of human knowledge is best viewed as mind-independent. Since, as noted, the notion of reality is intrinsically related to the idea of mind-independence, it follows that the concept of knowledge is also so related; in other words, the conception of human knowledge requires the realist position. Denial of reality—anti-realism—therefore, amounts to denial of knowledge: if there is no world, then there is no knowledge, for the world is the object of knowledge. Anti-realists, then, cannot uphold the idea of human knowledge. This consequence seriously dents the attractiveness of anti-realism. The two philosophical ideas that lead to this consequence are as follows.

The first of these ideas is commonly known as the justified true belief (JTB) conception of knowledge. The conception is traceable to Plato. In his allegory of the cave, Plato invites us to imagine a dark cave with some light filtering in from the outside through an opening. The light throws shadows of objects that pass by the opening, on a blank wall across the cave. There are people sitting chained between the source of light and the wall such that they can only see the shadows of objects, including of themselves, on the wall. They do not have any independent grasp of these objects. According to Plato, these people have at best some beliefs about the world, not knowledge.

Plato's Condition: For an epistemic subject S and a proposition p,

S knows that p just in case

(1) S believes that p,
(2) *p* is true,
(3) S is justified in believing that p

The first condition states that in order to attain knowledge that p, the subject S attains the state of belief that p; this condition relates the content of the proposition *p* to the epistemic subject. The second condition requires that only true propositions count; this condition relates the content of the proposition to truth. The third condition requires that S must have some evidence for (the truth of) p; this condition relates truth to the subject. Thus, the three conditions are said to be individually necessary and jointly sufficient for S's knowledge that p. There is wide agreement that these three conditions are needed at least to characterize knowledge. For the purpose of this chapter, I will hold on to this general agreement.

In a later chapter (Chap. 8), I will examine the JTB conception of knowledge in some detail to reject the idea that this conception of knowledge is a mental or psychological account of human knowledge. Also, following an influential paper by Edmund Gettier (1963), the Platonic conditions are viewed as inadequate since many counter-examples have been found that satisfy the three conditions without satisfying the intuitive conception of knowledge; hence philosophers have proposed various additional conditions. I am setting all these modifications aside since they do not affect the realism issue raised here.

The second of the two standard ideas in philosophy is what is known as the semantic definition of truth due to Alfred Tarski (1956).

Tarski Condition: For a sentence p in a language,

p is true if and only if p.
E.g., *Snow is white* is true if and only if snow is white.

According to Tarski, the truth-predicate *is true* exhibits the link between a sentence *p* of a language, and certain uses of the sentence that indicate the state of affairs mentioned in the sentence, the circumstance that p. In that sense, the truth-condition reflects the classical correspondence conception of truth: truth is a relation between language and the world. Thus, the sentence *snow is white* is true just in case snow, in fact, is white. This material effect is formally accomplished by naming, in the metalanguage, each sentence of an object language on the left-hand side of a material equivalence, and using a translation of the sentence in the metalanguage on the right-hand side. The significance of the definition becomes more perspicuous when, instead of a homophonic language such as English in which the object language is contained in the metalanguage, we mention a sentence of another language on the left-hand side:

Baraf safed hai is true if and only if snow is white.

There are controversies over whether Tarski's theory captures the classical correspondence theory of truth or whether it is merely a 'deflationary' account of truth (Strawson 1949; Horwich 2004). One could also dispute whether the conception of the world/states of affairs in correspondence theory necessarily signals a mind-independent world or merely an 'extra-linguistic' world. Elsewhere (Mukherji 2010), I have argued in detail against Donald Davidson's idea (Davidson 1967) that Tarski's condition be viewed as a theory of understanding of a sentence. Finally, one could invoke other conceptions of truth, such as coherence and pragmatic conceptions, to give an account of human knowledge.

Nevertheless, I will adhere to the assumptions of correspondence and mind-independent reality in the account of Tarski's condition not only because these are the most natural assumptions, but they also help bring out a powerful intuition about human knowledge. For example, it is unclear if other conceptions of truth can even be articulated without assuming the realist version of the correspondence theory of truth (Davidson 1990).

With this preparation, we may now use the Tarski equivalence to replace the expression '*p* is true' with 'p' in the second clause of JTB to obtain:

S knows that p just in case p, among other things.

Thus, S knows that snow is white when it is the case that snow is white. The formulation will be trivial if the whiteness of snow is a manifestation of S's mind; in effect, S will know that snow is white because S thinks so. That's exactly what Plato's condition was designed to rule out.

Such is the nature of human knowledge; human knowledge requires the conception of a mind-independent world. Our concept of knowledge begins to lose its regulative power if the object of knowledge fails to be a part of the *external* world; knowledge matters because the world does, every step of the way. A great deal of these powerful intuitions needs to be systematically rejected if we are to give up the proposed realist conception of knowledge. Suppose we hold on to the conception on that basis.

According to this conception of knowledge, which is based on strong philosophical foundations, alleged alternative conceptions of human knowledge as a mode of imagination, social construction, elaborate clan practice such as eurocentricity, expression of power in hegemonistic politics, a patriarchal trope since Plato, an instrument of control, an impediment to freedom, and the like, are simply beside the point. They do not affect the basic realist conception of human knowledge at all. On occasion, dimensions of human knowledge may be so used as to act, say, as an instrument of control. A glaring example is the knowledge of split atoms. Nuclear devices are instruments of control precisely because they exploit (realistic) knowledge of atoms. If atomic structure was a fiction, there would be no nuclear weapons.

2.3   The Design Problem

Despite the rather compelling conceptual connection between human knowledge and objective truth, the realist conception of knowledge needs to be severely qualified. We saw that the JTB conception gives an account of knowledge of a subject S. By now, it is well understood that the epistemic subject is an active participant in the attainment of knowledge. As stressed by the classical philosophers mentioned earlier, humans form a knowledge of the world by virtue of 'intelligible ideas and conceptions of things' from within human cognitive resources. The crucial point is that humans come to know the world by forming *conceptions* of things; these conceptions are largely human-dependent. In that sense, it is seriously questionable if the object of knowledge may be viewed as mind-independent.

In a classic paper titled 'What is it like to be a bat?', the philosopher Thomas Nagel (1974) suggested that, given fundamental differences in the sensory appa-ratus between humans and bats, humans can never experience the world as the bat does. By parity of reason, the human conception of the world is a product of its own perceptual resources, not available to the bats. Humans would have formed a very different conception of the world if they were endowed with the bat's equipment. Thus, our failure to adopt the bat's perspective is not the basic issue, for the bat's perspective is as species-specific as ours.

In any case, we know that humans perceive the world through a narrow band of, say, visual and auditory spectrums. For example, humans cannot directly perceive infra-red and ultra-sound effects, unless those effects are 'translated' within the range accessible to humans. Similarly, unlike many insects and other organisms, humans are unable to locate and identify objects with heat sensors, or find locations with geomagnetic tracking. What the world is like then is a product of the specific design of the epistemic subject.

In the next essay (Chap. 3) we will see that Immanuel Kant held that humans categorize their sensory experience of the world in terms of what he called 'productive imagination', a method of abstraction imposed by the mind. The task of productive imagination is to form schemata of objects given in experience by removing the 'excesses' of sensory information, and aligning the residual 'image' with categories already present in the mind. It is interesting that Kant also held that we possibly cannot have a 'theory' of the suggested alignment. We return to the issue repeatedly in this work, including later in this chapter.

In more recent work, the cognitive psychologist Elizabeth Spelke (2003) has shown that humans differ from other animals in their unique ability to combine information across broad categories such as colour and spatiality. Spelke (2010, p. 209) writes, 'the capacity to combine core concepts freely and productively may give humans a range of choice far beyond what either our learning history or our evolutionary history would seem to allow'. In recent studies, similar examples of human uniqueness from other domains of human knowledge abound (Penn et al. 2008). Many psychologists trace these abilities to the unique human endowment of language (Carey 2009).

Human knowledge is framed in the linguistic mode, as in JTB. As we saw, knowledge has the form (*that) snow is white*; that is how the objects and events of the world are organized for humans to have a knowledge of them. As Ludwig Wittgenstein (1922) would have said: <snow is white> is the thought, *snow is white* is what is uttered, and snow is white is what the world is like. It is not at all clear that nonlinguistic organisms categorize the world in familiar human terms at all. Analytically speaking, there is no doubt that all organisms organize their experiences with their own species-specific categories, but we may not have any idea what those categories are, as Nagel pointed out; worse, the 'organization' of their experience may not be categorial at all (Davidson 1975). So the troubling consequence is: what kind of world you know depends on what kind of species you are.

To summarize, on the one hand, we fondly entertain a conception of knowledge based on objective truth to achieve rationality in our living. Recall the justification clause in JTB: it is preferable to lead a life of knowledge than a (chained) life of beliefs. On the other, the very source of the coveted rationality, namely, the character of human design, threatens to turn human knowledge into a species-specific instinct, on par with the rest of the organisms. Given the rather severe boundaries of the problem, it is unclear how to retain the classical conception of knowledge.

The problem that human cognitive design poses for the conception of knowledge appears to be different from classical scepticism and some familiar versions of anti-realism. Scepticism denies the availability of knowledge as certain and truthful on the ground that human cognitive resources, such as the perceptual systems, are intrinsically unreliable and, hence, they may fail on occasion. Therefore, scepticism loses ground if perceptual and other systems function properly. For this reason, the general thrust against scepticism has been to argue that the idea that perceptual systems may fail globally is incoherent (Strawson 1985). In order to show that a perceptual system has failed in a certain case, we need to contrast it with a case where it didn't; global scepticism thus cannot be coherently articulated. In contrast, the design problem persists even if all cognitive resources function perfectly. In fact, the more perfectly they function, the more they impose human-specific 'conceptions of things' on human knowledge.

The basic thrust of anti-realism is to deny that there is a world for humans to have a knowledge of. For the anti-realist, human inquiry, especially abstract scientific inquiry, just provides some instrumental means of organizing sensory experience. It does not and cannot say what the world is like since the idea of a world beyond sensory experiences is a fiction. It is not clear that the design problem has the same effect. The problem no doubt stresses the central significance of human design in forming a conception of the world. But it does not strictly follow that the world so conceived is a fiction, because the conception of a fiction needs to be entertained within the resources of human design as well. It just follows that the world so conceived is intrinsically tied to human design. That is the only kind of world humans may conceive of with their kind of mind; nonhumans, with different cognitive apparatus, may conceive of other kinds of worlds. No coherent meaning can be attached to the idea that human knowledge gives an account of an otherwise mind-independent world. In the sense outlined earlier, the world is a human world.

The reassuring idea that the design problem does not totally rule out the concept of knowledge also helps in dispelling some standard responses in philosophy. At this level of generality involving the entirety of human knowledge, philosophical problems often acquire a cloudy, even mystical, character. The uncertain nature of those problems might encourage some strategy for a dissolution of the problem, say, by 'clarification', as Ludwig Wittgenstein suggested. In the present case, it could be argued that the problem seems to assume something that it rejects. To question the objectivity of knowledge on the basis of species-specificity of human design is to grant objective knowledge of that design, and the problem apparently stands defeated. Further, what do we mean when we say that nonhuman organisms conceive of other kinds of world? If we cannot ourselves conceive of any other kind of world except what we are allowed by our design, how can we make a comment on what differently-designed animals conceive of? As Chomsky (2001) suggests, Nagel's question, 'What is it like to be a bat?', does not seem to have an answer; hence, the question could be meaningless.

A short response to these suggestions is that it so happens that humans are simply endowed with the ability to reflect on their own design. Perhaps this endowment of self-reflection is a by-product of the linguistic endowment, but I'll leave that aside for now. No doubt such reflections are severely restricted by the very nature of the enterprise. As Thomas Nagel (1997) observed: 'There are inevitably going to be limits on the closure achievable by turning our procedures of understanding on themselves.' Within those restrictions, however, once we form a preliminary idea of how we engage with the world, the basis for the design problem is already laid, and the (mind-independent) world is progressively lost.

Holding on to that thought, it is tempting to infer that the design problem is a consequence of the very conception of knowledge. This is because, as the JTB formulation highlights, knowledge is ascribed to an (epistemic) agent. The design of the agent is an essential component of knowledge, and once we entertain such a concept of knowledge, the idea of mind-independent world begins to collapse. Therefore, even if there is some methodological merit in the 'clarificatory' moves suggested in the previous paragraph, they do not enable us to reclaim the lost mind-independent world. From our limited knowledge of human design, we know that humans can only *raise* Nagel's question, without answering it.

Some problems can only be contemplated in perplexity, if not in deafening silence. In fact, for the perplexity to arise, Nagel's specific question is not really needed. As Nagel himself observes, Nagel's question could be viewed as a rhetorical device—the method of difference, in this case between bats and humans—to highlight the design problem which persists even if humans are the only organisms around. In this chapter, I am setting aside the even more difficult issue that the design problem is not restricted to inter-species perspectives on the world, but generalizes to intra-species perspectives as well—between you and me. Solipsism signals the end of philosophy.

2.4   Forms of Scientific Realism

Although the problem is all pervasive, the design problem may have little effect on common life. Common life is typically intimately engaged with the phenomenal world to sort out the fake from the genuine in that world. So the further issue of whether what is taken to be genuine is (also) *real* might not appear in the consciousness to generate the required perplexity. The design problem becomes menacing in more reflective enterprises like literature, philosophy, and the sciences.

For thinkers like Chomsky (2000) and others, the scientific enterprise is geared to establish referentiality of its explanatory terms. And the concept of reference gets its salience from the overriding scientific norm that scientific thinking aims to discover real *joints of nature*, notwithstanding the use of elaborate instrumental means such as mathematical models, approximations, artificial experimental set-ups, etc. Thus, the idea that human knowledge and, thus, the conception of reality thereof, can only be products of human design raises some interesting issues for the coveted scientific realism.

According to authors like Karl Popper (1979), scientific realism is fundamentally an extension of classical realism: the world consists of entities that correspond to the concepts humans employ to form a view of the world. Realism assumes the existence of a mind-independent world, some of whose properties are correctly described by humans to attain knowledge of those aspects of the world. Modern science may be viewed as the most salient example of such knowledge. In view of the design problem, what sense can now be attached to the Popperian idea of realism?

In the context of, say, theoretical physics, scientific realism is discussed from two different directions, general and specific; in my view it is prudent to keep them separate. The first general issue concerns the referential terms of a scientific theory as noted: whether the terms employed in science actually pick out aspects/entities in the world. The problem became philosophically significant, beyond the original Platonic concerns, ever since modern science employed terms for unobservable entities and processes, such as electrons and gravitational fields, to explain the nature of the world. Even if we assume that our observational terms—*dog*, *mountain*—have real counterparts, it wasn't clear that the assumption may be extended to photons that don't have rest mass, not to mention dark matter that apparently fills much of the universe.

The second, more specific, issue concerned the character of quantum theory. Quantum theory postulated the unobserved state (wave function) of a microscopic system in terms of the Schrödinger equation. Whether the wave function is real then depends on whether its postulated effects may be observed. Here the 'uncertainty principle' poses a fundamental problem since the principle embodies the idea that some pairs of physical properties of a system, like the position and momentum of a particle, do not possess simultaneously precise values. According to one very prominent school of thought known as the Copenhagen interpretation, this implies that the 'objectively real' state of a physical system can never be precisely determined. This is because, in the very act of measurement of physical properties of the microscopic environment, the instruments of measurement, including the observer, interfere with the determination of these properties in an unavoidable way. It is important to emphasize that the fundamental problem is not just errors in measurement, it is the act of measurement itself.

It is a short step from this measurement problem to the anti-realist position that scientific description—in this case, the Schrödinger equation—is an instrument of representation of phenomenon rather than description of reality. Since the suggested anti-realism arises due to the unobservability of wave functions, quantum anti-realism also falls under the general problem of unobservability. But the crucial difference between quantum states of particles and such unobservable physical processes as gravitational and magnetic fields is that the unobservability in quantum theory arises due to the underlying uncertainty principle that endows physical systems with their non-deterministic statistical character.

Albert Einstein reacted to this peculiar consequence of quantum theory with his famous quip: 'God does not play dice with the universe.' According to Einstein, the problem is not that physical systems are uncertain; the problem is that quantum theory is an incomplete scientific theory. Thus, some physicists sought to recover the realist, deterministic picture of the physical universe by postulating 'hidden variables' allegedly missed in the quantum scheme. Other physicists since Einstein have attempted to 'recover' some notion of determinism with the many-worlds hypothesis, implicate order, and the like, the details of which need not concern us here. In this sense, quantum theory can be set aside without affecting the rest of physics. Therefore, the anti-realist picture posed by quantum theory does not extend to physics as such; physics might still satisfy the referential norm.

Nevertheless, once we fully engage with the design problem, it is not clear which notion of scientific realism survives given the obvious point that scientific theories cannot fail to be products of the human mind. To emphasize, both the realist view of quantum mechanics via hidden variables and the like, and the anti-realist, instrumentalist Copenhagen view are to be understood within the undeniable idea that quantum mechanics is a human product, just as relativity theory and string theory are. The specific realism issue concerning quantum theory has to do with the character of particular scientific theories, not with the identity of the species that constructs those theories; there is just one species. In that sense, even if physics is able to come up with a 'new physics' in which the measurement problem posed by quantum theory is solved (Penrose 2001; see Chap. 5), the new theory will continue to be affected by the design problem, on a par with otherwise realist theories such as Newtonian mechanics and relativity theory. So, the design problem forces the inquiry as to how some of the fundamental theories of the universe achieve the desired referential norm, even if we set the problematic quantum theory aside.

2.5 Nature and Order

In the next chapter (Chap. 3), following some suggestions due to Immanuel Kant and taking cues from the history of advanced sciences, we will see that some restricted notion of objective truth may still be available, within the bounds of human intelligibility, in some corners of human knowledge where we are able to invoke what is known as the Galilean style of inquiry. Here my goal is to briefly examine what effect the design problem has on the Galilean style.

It is of much interest that Galileo Galilei, the sixteenth-century Italian scientist often credited with laying the foundation of modern physics, actually held a rather grim view that humans will never completely understand even 'a single effect in nature'. We will examine the basis of his pessimism in some detail in the next chapter. For now, if science cannot explain 'a single effect in nature', how do we explain the sense of deep understanding, of genuine scientific explanation, in some selective domains? How is it that some instances of science, say, theoretical physics, convey an abiding sense of truth, a view of 'the real properties of the natural world'?

It is commonly held in science that the answer is located in the idea of mathematical physics. The mathematical formulae of physics are not only empirical in character, they also signal vast generalizations: Newtonian mechanics, relativity theory, quantum theory, and now string theory are often viewed as theories of everything. So, the really puzzling feature of the fundamental laws of physics is that they are at once mathematical in character and representations of large aspects of the universe. Thus, mathematical physics in the Galilean style stays within the limits of human intelligibility justifying Galileo's pessimism, but it raises the *standard* of intelligibility to a very high order to be able to pronounce discovery of the secrets of nature (see Chap. 3).

This primacy of mathematical physics as the ultimate form of human knowledge was recognized since the beginning of modern physics. Kepler (1609/1858) held that 'nature is always able to accomplish things through rather simple means, it doesn't act through difficult winding paths'. Galilei (1632) thought that 'nature generally employs only the least elaborate, the simplest and easiest of means... nature is perfect and simple, and creates nothing in vain'. Newton (1687) suggested that 'we are to admit no more causes to natural things than such as are both true and sufficient to explain their appearances... for nature is pleased with simplicity, and affects not the pomp of superfluous causes'. Einstein (1954) said that 'nature is the realization of the simplest conceivable mathematical ideas'.

Contemporary authors such as Weinberg (1976, 1993), in fact, trace the *realistic significance* of physics to its mathematical formulations: 'we have all been making abstract mathematical models of the universe to which at least the physicists give a higher degree of reality than they accord the ordinary world of sensations' (Weinberg 1976). Weinberg and others have called this form of explanation in physics the *Galilean Style* (Chomsky 1980). The style, according to these authors, works as a foundational methodological principle in science, especially physics.

The brief discussion of intelligibility and Galilean style brings out three salient aspects of the mathematized conception of reality.

(A) It abstracts away from the ordinary world of sensation; 'most of what we find around us in the world of ordinary experiences is unhelpful for determining the real properties of the natural world' (Chomsky 2000).
(B) It assumes the universe to have a 'simple', 'perfect' design because nothing else can be studied at the desired depth.
(C) It prioritizes mathematical models which are a priori and, hence, an exclusive product of the human mind.

Still, despite its impressive record, the Galilean style is available only rarely in human inquiry. Vast domains of ordinary human knowledge, thus, cannot ensure objective truth on this count. In any case, the Galilean style itself is a supreme example of human design, especially via (C): in a very significant sense, the Galilean style essentially constructs an imaginary world. Its late emergence in human history simply adds to the problem of design. So, by itself, the Galilean style does not seem to provide any definite analytical handle to meet the design problem.

Yet the amazing thing is that the Galilean style actually works! The world does seem to obey simple mathematical principles justifying the intellectual confidence of centuries of outstanding reflection, as we saw. Theoretical physics has unearthed deep secrets of nature in terms of strikingly simple laws; these can be understood only after elaborate mathematical argumentation. But there are other, more direct, examples. I will discuss two of the popular ones.

The Fibonacci sequence, allegedly discovered by the twelfth-century Italian mathematician Leonardo Fibonacci, is the series of numbers: 0, 1, 1, 2, 3, 5, 8, 13, 21, 34,... Starting with 0 and 1, the next number is found by adding the two numbers before it: 1 is $(0 + 1)$; 2 is $(1 + 1)$; 3 is $(1 + 2)$; 5 is $(2 + 3)$; and so on! The sequence is generated by a very simple *recursive* function that imposes the same order repeatedly at progressively higher levels: $x_n = x_{n-1} + x_{n-2}$. The series was also described by Indian mathematicians working on syllabic and metrical structure of speech perhaps much before Fibonacci did. Indologists suggest that 'the sequence Fn had already been discussed by Indian scholars, who had long been interested in rhythmic patterns... both Gopala (before 1135 AD) and Hemachandra (c. 1150) mentioned the numbers 1, 2, 3, 5, 8, 13, 21 explicitly' (Singh 1985).

The interesting aspect is that this progress of numbers maintains a constant ratio, the 'golden' ratio:

A	B	B/A (Golden ratio)
2	3	1.5
3	5	1.6666...
5	8	1.6, and so on.

When the numbers are plotted figuratively, the golden ratio generates a systematic spiral. The amazing thing is that the Fibonacci spiral is observed in many forms in nature, most notably in the organization of seeds, leaves and petals in flowers and other flora; they are also seen on snail-shells, starfish, bone structure of sharks, distribution of digits on human hands, etc. (see http://jwilson.coe.uga.edu/emat6680/parveen/fib_nature.htm).

The second, slightly more difficult example comes from the study of fractals, self-replicating geometrical shapes. A simple and small geometrical shape such as a triangle, or an open section of a curve, may be repeated at different levels and scales to generate some of the most complex phenomena in nature. The self-replication or self-similarity can come in a variety of forms: exact self-similarity as in some snowflakes; quasi self-similarity noted usually in artificial mathematical figures such as Mandelbrot sets; statistical self-similarity as in shape of coast-lines, and so on (see http://www.mnn.com/earth-matters/wilderness-resources/blogs/14-amazing-fractals-found-in-nature).

As with the Fibonacci sequence, fractals can be mathematically generated from simple recursive functions such as $z = z^2 + C$. From organization of leaves, snowflakes, patterns on snails, meandering rivers and coastlines, to solar systems, fractals exemplify Galileo's statement that 'nature generally employs only the least elaborate, the simplest and easiest of means... nature is perfect and simple, and creates nothing in vain'. The evidence is overwhelming.

In my view, such phenomena suggest a different angle on the design problem. The mathematical forms just cited—Fibonacci numbers and fractals—are no doubt contributions of the mind. But to observe their instances all over nature in astonishing variety does not seem to be occasional idealistic fantasy. It seems there are intimate connections between forms that arise in the human mind and the real forms of nature, and the connections are not satisfactorily explained by saying, 'oh, that's how our grasp of the world is designed by the mind'. The fit is so extensively varied and regular that it almost looks as if the mind-design is such that it simulates—perhaps, reciprocates—world-design; in these cases, Descartes' statue seems to just leap to the mind as a finished product. In other words, there is a strong intuition that the mind comes up with such a design because the world is so designed.

Several related caveats are immediately in order. First, it cannot be the case that each mind-design has a corresponding world-design. This is not only because sometimes even strong beliefs about the world are false, most mathematical forms conceived by the human mind are not realized in the world. Second, even if there is strong intuition about 'fit' in these cases, the intuition may well be aesthetic—based on simplicity, beauty, symmetry, harmony—rather than on truth. Third, given the severe restrictions on inquiry on these questions imposed by the design problem, it is hard to see how to make analytical and empirical progress on the question of 'fit'; there is no 'third party'.

Even with these caveats, it is simply counter-intuitive to ascribe the strong sense of fit to 'idealistic fantasy'. A disclaimer of the form 'it is nothing but mind-design' just does not explain the strength of the 'externalist' intuition when the curvature of the coastline or the arrangement of sunflower seeds unfold, as we zoom in; this intuition could be the primary source of mathematical realism, but I will set it aside. In general, as the tortuous discussion back and forth between realism and anti-realism in this chapter suggests, the idea that the world is something of a fiction is deeply problematic. The philosopher Paul Edwards once suggested that, no matter how many sunsets, rushing rivers and naked women you show, the idealist will keep on saying, 'further manifestations of my mind'. The remark is hilarious precisely because idealism is counter-intuitive.

Yet, as the caveats suggest, there does not seem to be any 'objective' route available to inquire into and establish the fit, and the mind-independent world continues to be elusive. It looks like the end of philosophy. I am aware that, following the work of Hilary Putnam and others, the literature in analytical metaphysics offers some other complicated alternatives, such as 'internal realism', at this point. My general feeling is that they ultimately fail to meet Nagel's stricture about limits on the closure achievable when human inquiry turns on itself.

Be that as it may, could it be that a very different style of inquiry might alleviate some of the discomfort with the preceding picture? Could it be that, in looking for a mind-independent world 'out there', we are adopting the wrong perspective on the problem? Maybe we are unable to locate the world because, in our analytical mode, we are *looking* for it. Perhaps there is nothing to look *for*, but everything to live *in*. I am reminded of the Zen master's quip when the disciple asked in frustration if the fly will ever be released from the bottle. The master replied that there was nothing to worry if the food was good and well-eaten, and the utensils were clean: the fly was out. Are we missing the world in our anxiety to grasp it?

To pursue the thought, let me collect some of the points mentioned mostly in passing earlier: (a) the problem of realism is not likely to occupy consciousness in common life because that life is immersed in the phenomenal world, the world of experience; (b) the realism issue *significantly* arises only with unobservable entities and processes; dogs and mountains certify our robust sense of reality when other things are equal; (c) the sense of reality could be an aesthetic consequence of certain stretches of experience, rather than an analytical move to account for those experiences.

2.6   The Table and the Poet

During his conversation with the physicist Albert Einstein, the poet-composer Rabindranath Tagore stated that 'truth, which is one with the universal being, must be essentially human; otherwise, whatever we individuals realize as true, never can be called truth'. 'The truth which is described as scientific', he continued, 'can (only) be reached through the process of logic' which itself is 'an organ of thought which is human' (in Marianoff 1930).

The point of interest in these remarks is that Tagore is concerned with truths of physics, scientific truth. For him, scientific truth is one with universal being; in other words, truth is universal. While claiming that truth is 'human', there is no emphasis on social norms, cultural forms, historicity, artistic variation, etc. In particular, he is not proposing that truth is subjective, whatever that problematic notion amounts to. In that sense, truth is objective. He seems to be upholding the concept of objective truth *while* asserting its essential humanness; in fact, he suggests that humanness is a *necessary* condition for objective truth. In our terms, it means that objective truth is not only consistent with human design, truth and design complement each other.

Tagore was a great poet and a humanist. As a humanist thinker, he also lectured and wrote on a variety of topics of general human interest such as the character of human existence, the play of nature on human creativity, and the role of values and religion in human societies. In one broad common sense of the term *philosophy*, the writings just mentioned may well be characterized as philosophical insofar as they are general reflections on the human condition. However, these reflective works are not philosophical in the narrower, more academic sense. Broadly viewed as systematic reflections on the nature of language, thought and reality and the relations between them, academic philosophy can be safely identified, as with other academic disciplines, with its textual lineage (Mukherji 2002, 2005).

Tagore's views on physics and science, therefore, need to be viewed not as technical comments, but in terms of general humanist reflections of a literary mind. Even then Tagore's 'outsider' view carries much intellectual interest. A work of art, including the writing of poetry, is not merely a play of form. Somehow the artist has to relate the emerging forms in his aesthetic imagination to stable aspects of human experience for the forms to have an irresistible interpretation. In that sense, an artist constantly struggles with the elusive reality of artistic depictions, giving rise to the other formidable issue of realism in the arts. An artist and poet of Tagore's genius was likely to have reached a satisfactory reflective understanding of his own artistic expressions that might have interesting philosophical implications. Elsewhere (Mukherji 2012), I have discussed how Tagore's poetic contemplation of bird-songs illuminates an empirically-viable conception of human musical ability. I wish to adopt a similar strategy for Tagore's 'poetic' views on science and truth.

Albert Einstein also was not an academic philosopher in the sense outlined above. However, his philosophical location with respect to the issue of scientific realism was far more intimate than Tagore's, as we saw. Thus, both for his outstanding role as a practicing scientist and a reflective thinker on the nature of science—especially, on the new physics—Einstein's remarks on the nature of reality carry intrinsic significance.

The Tagore–Einstein conversation does not really have the form of a debate. Rather, it has the form of repeated assertions by two minds reflecting in parallel. And it is not very clear what these assertions were about. For example, from some parts of the conversation it appears that the conversation was about whether there is a mind-independent reality. Dmitri Marianoff, who reported these conversations in the *New York Times*, titled the first dialogue, 'Thoughts on the possibility of [truth's] Existence without relation to Humanity'. It is suggested that Einstein says 'yes', Tagore says 'no'. It is unclear that there was in fact such a direct opposition.

Consider the issue of the table. While Einstein clearly holds that the table continues to be there even if no one sees it, Tagore does not quite say that the table won't be there. He says instead that it will be there but under the gaze of a universal mind. So Tagore does not deny the existence of the table when no individual human is present to perceive it; he ascribes the existence in that case to the presence of a universal mind. Similarly, in his *Three Dialogues between Hylas and Philonous*, George Berkeley held that since sensible things may exist independently of human beings, 'there must be some other mind wherein they exist' (Berkeley 1731). For Berkeley, this 'other' mind is God; Tagore called it variously 'supreme man,' 'universal being,' and the like.

The standard criticisms against Berkeley thus apply to Tagore as well. The postulation of a constantly and universally aware supreme mind does not seem to have more explanatory power than the simpler postulation of the (existing) table itself. The universal mind will need the table for a veridical perception of it in any case; if that perception is non-veridical, the table won't be there which is a consequence both Berkeley and Tagore deny. So, Tagore needs to postulate all of: (a) a universal mind in the form of all-pervasive consciousness; (b) the ability of the human mind to grasp the universal mind; *and* (c) veridical perception of the table by the universal mind, just to say that the table continues to be there. Even though Einstein does not offer any further realist argument in favour of his view that the table continues to be there ('I cannot prove my conception is right, but that is my religion'), his robust and simple common sense seems to outweigh Tagore's complicated idealist thrust.

Philosophical difficulties aside, for the case under study, humans do seem to have a preference for the simpler option of settling for the reality of the table because that sense of reality is in fact withdrawn in some special cases. For one, even if, other things being equal, we robustly believe that the table continues to exist when I leave the room, we do not believe that my shadow—which also I can see among other 'thing'-like stuff—will continue to exist in the room if I leave the room. This is because other things are not equal; unlike the table, the continued existence of my shadow is tied to my presence in the room. So, there is a clear distinction between things to which we do or do not ascribe (independent) reality even if they are perceptible in a non-illusory way. Note that, at this stage, I am not proposing that the design problem thus stands refuted by common consensus; that cannot be the case, as we saw. All that I am doing now is to inquire into the character of human commonsense to see why the design problem fails to have a near-fatal grip there.

To reject Tagore's idealistic argument is not to deny the appeal of Tagore's basic assertion that 'the truth which is described as scientific and which only can be reached through the process of logic—in other words, by an organ of thought which is human'. However, Tagore's own explanation of what he means is not very helpful, as we saw. For Tagore, scientific truth obtains when the human mind forms a 'universal harmony' with the 'supreme being, *Brahman*'. Such an explanation seems unnecessary because the basic claim is obvious as scientific theories are unfailingly human products. Furthermore, the postulation of a universal being takes the explanation away from the *individual* human being whose perception of the table and the conception of scientific truth *are* the relevant phenomena. In that sense, Tagore in fact loses his grip on the 'humanness' of the issue.

Earlier, I mentioned the reflective value of Tagore's poetic conception of how things are such as bird-songs. Naturally, then, Tagore's deeply reflective views on the nature of reality is likely to be represented more effectively in his literary work, rather than in occasional philosophical pronouncements of uncertain value as we saw. Consider the first two lines of his very popular lyric *mahāvishwe mahākāshe* (my translation from Bengali throughout):

> In the cosmos, the endless sky, the boundless time
>
> I, the human, travel alone in wonder, in wonder.

The first line depicts what philosophers call the 'manifold' consisting of all that is there in the universe located in the universal space-time framework. The second line 'humanizes' this manifold. I have access to the *entire* manifold for further reflection because I can be *there* at every possible point of this manifold as a part of my travel itinerary. Tagore emphasizes the idea of traveling *in wonder*; he mentions it twice in succession. Perhaps he means that, if the manifold were to be my (subjective) *construction*, an effect of my conscious imagination, then there is no travel, no wonder. I travel in wonder because the manifold is there independently of my reflection such that I *experience* the manifold just as I experience the table out there.

Now, it cannot be literally correct that the entire manifold is open to my experience; I can possibly cover only a tiny fragment of the manifold. Yet the point of wonder will be missed if I think of the unexperienced parts of the manifold either as a manifestation of my imagination or as a world unknowable. For the poet, this *impasse* does not arise because the manifold is not contemplated in isolation of my travels. In a sense, my travels reveal the manifold already there; the unrevealed manifold must already also be there to facilitate the conditions for my continued travel. As noted earlier, the form of discourse, loaded heavily with metaphors, is not analytic in character. It is rather an expression of convictions, perhaps even a dynamic report, as the subject's experiences uncover further aspects of the universe.

At many places in his creative work, Tagore elaborates on the notion of a cosmos which unfolds due to human intervention. Consider another very popular lyric *ākāshbharā suryatārā*.

> The sky is full of the sun and stars, the world is full of life
>
> I have found my abode there,
>
> Thus my song comes alive in wonder.
>
> The waves of eternal time that cause the ebb and tide
>
> Also guide the rush of blood through my veins,
>
> Thus my song comes alive in wonder.
>
> When I walk on the grass on the forest-path,
>
> The scent of the flowers startle my spirit
>
> The gift of joy is scattered all around
>
> Thus my song comes alive in wonder...

In line 1, Tagore begins to describe the cosmos as a skyful of sun and stars, but it also includes the life in the universe. In line 2, just as in the earlier poem, he locates the I, the poet himself, amongst the elements of the cosmos. Line 3 says, this wonder of self-realized cosmos gives rise to the poet's song. In lines 4 and 5, Tagore highlights the unity of the cosmic order by noting that the reason why my blood flows is the same as ebb following tide for eternity. In lines 7 and 8, the idea of the cosmos is expanded to include blades of grass on the forest path (7) and the smell of flowers (8), both giving rise to the joy and wonder (lines 6 and 9). Similar sentiments can be found in dozens of other poems and songs.

How do we interpret this image of a human universe? In some literary and philosophical circles, Tagore's colourful 'animated' conception of the cosmos is often overinterpreted to suggest parallels with the *Upanishadik* tradition, or the primacy of artistic (fictional) imagination. In both cases, the emphasis is placed on the subject, the I, as *constructing* the cosmos. In other words, the cosmos is viewed as *dependent* on the solitary self, the cosmos *disappears* if the self does. Tagore is thus viewed as a classical idealist. As noted, some of his inadequate philosophical arguments do suggest such a picture, wrongly in my view.

Tagore, as noted, did not develop his complex worldview with professional philosophical rigour. There is no doubt that his conception of the universe is a fallout of his artistic endeavours; it is not a systematic product of cold analytical reflection. Also, Tagore's conception of universe is scattered in a large body of artistic work whose meaning changed sharply throughout his life; in some cases, he did leave the impression of an idealist's universe. For example in one of his much-cited poems, he seems to assert that the emerald gets its colour due to my consciousness, the sky is lit up when I open my eyes, etc. Authors cite these lines with glee to emphasize Tagore's 'subjectivism' and affinity with quantum theory (Ghose 2010, pp. cxli–cl).

But it is routinely missed in the literature that, in this poem, Tagore carefully keeps to what philosophers call 'secondary qualities' after the seventeenth-century British philosopher John Locke—qualities of matter that seem to manifest only under human observation and are not intrinsic to matter. Tagore also mentions aesthetic properties such as 'beautiful' which, of course, have intrinsic reference to the beholder. Yet there is no evidence that Tagore thought that the so-called 'primary qualities' such as size and density of objects—proper subject matters of physics—are also products of the beholder.

Be that as it may, let us assume that Tagore might not have been fully consistent in his worldview through his varied artistic expression. Yet, as explained in some detail above, a central part of his artistic work seems to project the conception of a human reality which is metaphysically independent of the human mind, but gets its entire epistemic significance via human experience. This is the sense in which the universe is at once familiar and unfamiliar (*ciradivaser vishwa ankhisammukhei*):

The perennial universe in front of my eyes,

I have seen a thousand times

At my door.

This timeless familiarity of the unfamiliar

Has filled the deep recesses of my heart

At ease.

So, the human universe is an *experienced* universe; only in human experience does the universe carry its full metaphysical significance. The universe appears to us in all its variegated complexity because humans—only humans—are endowed with 'organs of thought' that transform the experience into a symbolic conception. And an experienced universe has the self as the interpreter of experience at its cognitive center. As Tagore puts it (*mānuSher ahankār patei*): the grand-designer works on the canvas of human ego.

To emphasize, fractals and Fibonacci sequences, among many other similar recursive forms, are products of the ingenuous human (mathematical) mind and, as noted, mathematical ideas, by themselves, are entirely a priori, that is, they are independent of the sensibilia. Yet, these beautiful 'fictions' seem to fill all parts of nature. It is in this specific sense of the mathematization of reality that we may understand Tagore's poetic idea that 'the human mind forms a universal harmony with the supreme being', except that nature itself is that supreme being.

Strictly speaking, the artistic conception of harmony with the universe does not really address the design problem in its original formulation; nothing can, as we saw. As in the philosophy of Immanuel Kant and later in the work of continental authors, Tagore placed the emphasis, not so much on the immanent world, as on the primacy of human experience. As we will see in the next chapter, Kant held that any rational account of human experience demands some conception of a mind-independent reality because, otherwise, we are left with no conception of what these experiences are experiences *of* (see Mukherjee 2007 for Tagore's familiarity with Kant). A poet's perspective goes beyond a mere postulation of the world to a direct 'travel' in it in wonder since the world *is* what human consciousness experiences. Following the analytic route, we end up with posing human design as a *problem* and lose the world as a consequence. In the poet's aesthetic route, the design is the source of wonder as human artistic imagination enables the world to unfold in front of human consciousness.

Deep scientific inquiry into the order of nature, as in hundreds of years of mathematical physics, requires even more ingenuity, not only to experience nature, say, in terms of Fibonacci numbers, but to explain the inner structure of matter that gives rise to its surface systematic forms. That deeper inquiry, resulting in the postulation and discovery of fundamental laws governing basic elements and forces can only be 'reached through the process of logic which itself is an organ of thought which is human,' as Tagore emphasized in his poetic gesture. There is no fundamental divide between scientific and artistic forms of inquiry. Humans, after exercise of ingenuity in either form of inquiry, come to grasp reality because reality, in that explicit sense, is human.

References

Berkeley, G. 1731. *Three Dialogues between Hylas and Philonous*. In *The Principles of Human Knowledge*, G. J. Warnock (Ed.). London: Collins/Fontana. 1962.

Carey, S. 2009. *Origin of Concepts*. New York: Oxford University Press.

Chomsky, N. 1980. *Rules and Representations*. Oxford: Basil Blackwell.

Chomsky, N. 2000. *New Horizons in the Study on Language and Mind*. Cambridge: Cambridge University Press.

Chomsky, N. 2001. Language and the rest of the world. In *Bose Memorial Lecture in Philosophy*, Delhi University, November 4.

Cudworth, R. 1731/1996. *A Treatise Concerning Eternal and Immutable Morality*. In *Cambridge Texts in the History of Philosophy*, S. Hutton (Ed.). Cambridge: Cambridge University Press. 1996.

Davidson, D. 1967. Truth and meaning. *Synthese* 17: 304–323.

Davidson, D. 1975. Thought and talk. In *Mind and Language*, S. Guttenplan (Ed.). Oxford: Oxford University Press.

Davidson, D. 1990. The structure and content of truth (The Dewey Lectures 1989). *Journal of Philosophy* 87: 279–328.

Einstein, A. 1954. *Ideas and Opinions*. New York: Bonanza Books.

Galilei, G. 1632. *Dialogue Concerning the Two Chief World Systems*. Berkeley: University of California Press. 1962.

Gettier, E. 1963. Is justified true belief knowledge? *Analysis* 23: 121–123.

Ghose, P. 2010. Introduction. In *Materialism and Immaterialism in India and the West: Varying Vistas*, P. Ghose (Ed.). New Delhi: Center for Studies in Civilizations.

Horwich, P. 2004. *From a Deflationary Point of View*. New York: Oxford University Press.

Kepler, J. 1609. *Astronomia nova* (New Astronomy).

Marianoff, D. 1930. Einstein and Tagore Plumb the truth: Scientist and Poet Exchange Thoughts on the possibility of its Existence without relation to Humanity. *New York Times*, 10 August.

Mukherjee, B. 2007. *Darshonik Dwijendranath*. In Bengali. *Akademi Patrika*, 22, 9–48. Kolkata: Bangla Akademi.

Mukherji, N. 2002. Academic philosophy in India. *Economic and Political Weekly* 37, 10 March, 931–36.

Mukherji, N. 2005. Textuality and common life. In *Literature and Philosophy: Essaying connections*. S. Chaudhury (Ed.). Kolkata: Papyrus and Jadavpur University.

Mukherji, N. 2010. *The Primacy of Grammar*. Cambridge: MIT Press.

Mukherji, N. 2012. I sing my song. In Bengali (*aami gaai gaan*). *Anustup*, Festival Number (*Sharodiya Sankhya*).

Nagel, T. 1974. What is it like to be a bat? *Philosophical Review* LXXXIII (4): 435–50.

Nagel, T. 1997. *The Last Word*. New York: Oxford University Press.

Newton, I. 1687. *Philosophie Naturalis Principia Mathematica*. Translated by Andrew Motte as *Newton's Principia: The Mathematical Principles Of Natural Philosophy* (1846). London: Kessinger Publishing Co.

Penn, D., K. Holyoak, and D. Povinelli. 2008. Darwin's mistake: Explaining the discontinuity between human and nonhuman minds. *Behavioral and Brain Sciences* 31: 109–178.

Penrose, R.. 2001. Conversation. In *Mind, Matter and Mystery: Questions in Science and Philosophy*, R. Nair (Ed.), 119–133. New Delhi: Scientia.

Popper, K.R. 1979. *Objective Knowledge: An Evolutionary Approach*. Oxford: Oxford University Press.

Singh, P. 1985. The so-called Fibonacci numbers in ancient and medieval India. *Historia Mathematica* 12 (3): 229–244.

Spelke, E. 2003. What makes us smart? Core knowledge and natural language. In *Language in Mind: Advances in the Investigation of Language and Thought*, D. Gentner, and S. Goldin-Meadow (Eds.). Cambridge, MA: MIT Press.

Spelke, E. 2010. Innateness, choice, and language. In *Chomsky Notebook*, J. Bricmont, and J. Franck (Eds.), 203–210. New York: Columbia University Press.

Strawson, P. 1949. Truth. *Analysis*, 9, 6 June, 83–97.

Strawson, P. 1985. *Skepticism and Naturalism: Some Varieties*. New York: Columbia University Press.

Tarski, A. 1956. Semantic conception of truth. In *Logic, Semantics, Metamathematics*, A. Tarski (Ed.). Translated by J. Woodger, 152–278. Oxford: Clarendon Press.

Weinberg, S. 1976. The forces of nature. *Bulletin of the American Academy of Arts and Sciences* 29: 13–29.

Weinberg, S. 1993. *Dreams of a Final Theory*. New York: Vintage.

Wittgenstein, L. 1922. *Tractatus Logico-Philosophicus*. Translated by D. Pears, and B. McGuiness. London: Routledge and Kegan Paul.

Chapter 3
Science and the Mind

> Science is a very strange activity. It only works for simple
> problems. Even in the hard sciences, when you move beyond the
> simplest structures, it becomes very descriptive.
>
> Noam Chomsky

As a fundamentally new mode of human inquiry, the development of mathematical physics several centuries ago is often characterized as the Newtonian revolution. The emergence of cognitive science in recent decades—the new science of the mind —is arguably as dramatic a phenomenon in human history as the advent of Newtonian mechanics. While Newtonian physics precisely aligned human reflective ingenuity with deep mysteries of the external world, the new science of the mind aims to do the same for the 'inner' world. If the programme is broadly successful, it will finally bring several millennia of philosophical and artistic exploration of the inner world under scientific control. Setting aside the issue of whether such a programme is (humanly) desirable, what does it mean for such a science to come into being in the first place?

There is an old adage that a theory of mind is an impossibility since the theory itself will be a product of the mind, and hence a part of the object under examination. The adage appeals to the image of eyeglasses: we can give only a partial and distorted description of the glasses when we wear them; we can take them off, but then we cannot see. This adage is distinct from classical scepticism that denies the possibility of any knowledge. The effect of the adage is restricted only to cognitive inquiry; in that sense, it allows the possibility of knowledge of the external world, say, the world of physics (Mukherji 2010, Chap. 1 for more).

This is a revised version of an earlier paper published as Mukherji (2009).

© Springer Nature Singapore Pte Ltd 2017
N. Mukherji, *Reflections on Human Inquiry*,
DOI 10.1007/978-981-10-5364-1_3

3.1   The Realism Issue

What, then, are the prospects for this new science of the mind when seen in the light of the more advanced sciences? An answer to this question requires at least a rough idea of the advanced sciences: what are the general principles that guide theoretical inquiry in an advanced science such as physics? The discipline of philosophy of science consists of a very large literature on this topic covering a number of classical debates on the character of the sciences. For my limited goals in this chapter, I will basically stay away from this technical literature and concentrate on some of the views of Noam Chomsky and Immanuel Kant for reasons that follow.

Chomsky remarks that one of the most compelling images of modern science is its apparent ability to unearth the 'real properties of matter' from below the chaos encountered in the 'world of ordinary experiences' (Chomsky 1991). The image appears to support a common conception of science that views it as heralding truth: the realist position. On this view, science tells us what basic objects underlie the constitution of the world and what are the principles governing them. The realist view explains the deep power and attraction of modern science; it also explains the almost fanatic motivation governing scientific practice.

In contrast, there is a minority view in the philosophy of science which says that scientific theories, including the most telling achievements of mathematical physics, are just figments of the human mind such that they necessarily fail to describe the (real) world as-it-is: the anti-realist position (see also Chap. 2 in this volume). In its extreme form, held notably by Nancy Cartwright (1983), this view says that scientific theories are *lies*. As noted, there is an astronomical literature on this debate, including involved discussion on exactly how to formulate the realism issue. For my purposes, the rough and ready view just sketched suffices.

As the leading exponent of one of the principal branches of cognitive science, namely, theoretical linguistics (also known as *biolinguistics*), Chomsky is likely to believe that the field of linguistics promises to meet the realist criterion of science; Chomsky is unlikely to hold that human language has such and such properties because he thinks so. For example, in the same paper cited above (Chomsky 1991), he says that 'rather far-reaching results' have been achieved in the domain of language. The results have been 'far-reaching' in the sense that something has been learned in these domains at a sufficient remove from 'the world of ordinary experiences'. In that sense, biolinguistics is arguably the only attempt in the history of ideas in which a study of an aspect of the human mind—language—is beginning to have the 'feel of scientific inquiry' (Chomsky, cited in Piattelli-Palmarini 1998). So, Chomsky needs the realist picture of science in general to promote the specific realist picture for the science of language.

Yet, as we will see in the next section, Chomsky also shares a largely anti-realist view for much of the history of science. More alarmingly, his anti-realism appears to be most prominent for the realm of science—namely, the history of physics beginning with Newton—to which some version of realism most effectively applies. So, what's going on? In this chapter, I attempt a reconciliation of the contrasting views. I suggest that a realist view of science does hold via what is known as the 'Galilean style' , but it holds very rarely in the history of the sciences and, when it does, it obtains for very restricted domains in those sciences. Much of what falls under the label *science*, then, does admit of possible anti-realist views.

I am aware that this rather exalted and demanding view of science can be contested. Many philosophers and sociologists of science may wish to lower the bar significantly. This is not the place to enter into a full discussion of the character of science. Thus, suppose we adopt the suggested view of science via the Galilean style for the purposes of this chapter. If the more advanced sciences are such a rarity in human inquiry, can we really expect a genuine science of the mind on par with theoretical physics? To emphasize, contrary to widespread anti-realist   views prevalent in humanities and social sciences, it is not denied that science does in fact achieve its realist goal in some rare, historically favourable cases. The question is: what does it mean for inquiries in cognitive science to fall in the same class, if at all?

At a first glance, it seems that the stated goal of cognitive science cannot be reached in practice. Although the broad discipline of cognitive science does propose to study the mind (Wilson and Keil 1999, p. xiii), the received notion of mind seems to be too heterogeneous to admit the Galilean style. It is hard to see how cognitive science is going to extract and articulate the 'real properties' of the mind from the obvious complexity of mental phenomena encountered in the 'world of ordinary experiences', as we observe the cognitive behaviour of a wide range of organisms. Even the least developed organisms such as bacteria and viruses are endowed with some form of sensory systems to grasp relevant parts of the world. As we go up the evolutionary ladder, most organisms show signs of consciousness and some ability for goal-directed behaviour, often planned in advance. The phenomenon explodes in diversity with the emergence of mammals that finally led to primates and the hominid line.

It is not surprising, then, that, as the cognitive sciences expanded, more and more phenomena became visible while the prospects for a (unified) theory of mind receded. In his introduction to *An Invitation to Cognitive Science*, Daniel Osherson observes that the topics covered in the four-volume series 'range from muscle movement to human rationality, from acoustic phonetics to mental imagery, from the cerebral locus of language to the categories that people use to organize experience' (1995, p. ix). Osherson also remarks that 'cognitive science is the study of human intelligence in all its forms, from perception and action to language and reasoning' (Osherson 1995, p. xi). Cognitive science, Osherson suggests, is concerned with 'everything we do', where 'we' also includes other organisms. Osherson concludes that 'topics as diverse as these require distinctive kinds of theories, tested against distinctive kinds of data'.

Prima facie, the task for a theory of mind is to furnish some general principles underlying this vast phenomenal range. The goal seems initially elusive since the basic sciences like physics and chemistry, that have unearthed simple, unified principles of nature, have nothing to say about mental phenomena. As a result, the apparently unordered vastness of cognitive phenomena is in a way forced on us in the absence of theoretical guidance from the basic sciences. As Chomsky (2000, p. 104) remarks, any 'complex system will appear to be a hopeless array of confusion before it comes to be understood, and its principles of organization and function discovered'.

There is a perceptible conflict, therefore, between the lofty goals of a unified science of the mind and the current state of the cognitive sciences. It is instructive at this point to reflect on what lessons can be learnt from the history of the more basic sciences on the stated goal of extraction of unified theories. What is involved in the adoption of the Galilean style such that some real properties of matter begin to show up from below the vast complexity of common experience? If we have a rough idea, then we may inquire if there are aspects of the otherwise complex mental phenomena where something like the Galilean style may be attainable.

3.2 Restriction to Intelligibility

As noted, Chomsky himself suggests an anti-realist position on science in some other writings. According to him, lessons from the history of natural sciences seem to suggest that 'most things cannot be studied by contemporary science' (Chomsky 2001; Hinzen 2006). On this issue, it seems to him that Galileo's intuition that humans will never completely understand even 'a single effect in nature' is more plausible than Descartes' confidence that 'most of the phenomena of nature could be explained in mechanical terms: the inorganic and organic world apart from humans, but also human physiology, sensation, perception, and action'. Developments in post-Cartesian science, especially Newtonian science, Chomsky holds, 'not only effectively destroyed the entire materialist, physicalist conception of the universe, but also *the standards of intelligibility* that were based on it' (emphasis added). Thus, Chomsky supports Alexander Koyre's remark that 'we simply have to accept that the world is constituted of entities and processes that we cannot intuitively grasp' (cited in Chomsky 2001).

The force of the expression *intuitively* seems to be that we cannot have direct knowledge of how the world is like; the knowledge has to be routed in terms of the resources available to our theory-building abilities. Thus, any conception of the universe is restricted to what is intelligible to us: as standards of intelligibility fall, so does our grasp of the universe. The restriction gives rise to the old irony that the world which undoubtedly gives rise to our knowledge of it cannot be sufficiently grasped by the only means available to us. Clearly, unlike the adage about spectacles mentioned above, these remarks are meant to apply to the whole of science including the most ingenuous proposals in theoretical physics.

In view of such severe restrictions on the possibility of reaching genuine theoretical understanding, it is unsurprising that it affects scientific pursuits like biology somewhat more directly than physics. Biological systems are not only immensely complex, they are commonly viewed as poor solutions to the design-problems posed by nature. These are, as Chomsky puts it, 'the best solution that evolution could achieve under existing circumstances, but perhaps a clumsy and messy solution' (2000, p. 18). Given the complexity and apparent clumsiness of biological systems it is difficult to achieve theoretical abstractions beyond systematic description. According to Chomsky, the study even of the 'lower form lies beyond the reach of theoretical understanding'.

Consider, for example, the research on nematodes, a very simple organism with no more than a few hundred neurons; so, people have been able to chart out their wiring diagrams and developmental patterns fairly accurately. Yet Chomsky reports that an entire research group at MIT devoted to the study of 'the stupid little worm' could not figure out why the 'worm does the things it does' (Chomsky 1994). These remarks have an obvious bearing on the biological system at issue, namely, the human mind. 'Chomsky has argued', the philosopher Daniel Dennett complains, 'that science has limits and, in particular, it stubs its toe on the mind' (Dennett 1995, pp. 386–7). We will return to Dennett's comment from other directions elsewhere in this volume (see 'Literature and Common Life', Chap. 11 in this volume).

Given the inverse relationship between complex systems and our understanding of them, it is even more likely that most other domains of human inquiry, especially those concerned with aspects of the human condition itself, will fail to yield any theoretical understanding at all:

> The idea that deep scientific analysis tells you something about problems of human beings and our lives and our interrelations with one another and so on is mostly pretence in my opinion—self-serving pretence, which is itself a technique of domination and exploitation (Chomsky 2000, p. 2).

Nevertheless, we are still left with at least theoretical physics, and it seems contemporary linguistics, where, to cite Chomsky again, 'far-reaching' results have been reached at a sufficient remove from 'the world of ordinary experiences'. This domain of cognitive psychology, Chomsky holds, has enabled the adoption of Galilean style so far achieved only in physics and a few other basic sciences. If science cannot explain 'a single effect in nature', how do we explain the sense of deep understanding, of genuine scientific explanation, in these selective domains? How is it that some instances of science, theoretical physics and linguistics, convey an abiding sense of truth, a view of 'the real properties of the natural world'?

The interest is that even anti-realism is required to explain the stark asymmetry in the depth of scientific explanation between, say, theoretical physics (plus linguistics) on the one hand, and meteorology and biology on the other. The 'sense of truth' that is felt in theoretical physics is simply not located in the explanatory format of, say, evolutionary biology. A natural thought is to try to examine how the mind, by virtue of its cognoscitive powers, can apply abstractions progressively on the complexity of experiences to reach simpler conceptions of the world on favourable occasions such as the domain of physics while failing to apply to domains of biology, meteorology, and the like. To that end, I will examine some proposals due to Immanuel Kant shortly.

To return to the realism debate, I find both the apparently opposing—realist and anti-realist—views to be intrinsically compelling. As cited, Chomsky himself seems to thrive on this tension: he claims at once that we may not understand a single effect in nature as well as that contemporary linguistics might have unearthed some real properties of nature. In my view, the source of the tension is that, as a new science, biolinguistics initiated by him remains isolated from the rest of the established sciences, especially biology (Mukherji 2010, Chap. 1). Not surprisingly, Chomsky (1995, pp. 1–2) places the burden on biology itself rather than on the scientific stature of linguistics:

> (H)ow can a system such as human language arise in the mind/brain, or for that matter, in the organic world, in which one seems not to find anything like the basic properties of human language? The concerns are appropriate, but their locus is misplaced; they are primarily a problem for biology and the brain sciences, which, as currently understood, do not provide any basis for what appear to be fairly well established conclusions about language.

So, there is a need to promote a notion of non-reductive scientific inquiry that stands on its own (Hinzen 2006; Mukherji 2010). The claim for the (advanced) scientific character of biolinguistics then has to be maintained without the advantage of support from the 'basic sciences'. A natural way of upholding this claim is to deny that the basic sciences have any more claim to truth than biolinguistics. Given the anti-realist conception of science, the demand for reduction to a basic science loses force. However, Chomsky does not wish this anti-realist move to cast doubt on the scientific character of biolinguistics itself; hence, the suggested feel of 'genuine scientific inquiry' perhaps leading to the unearthing of 'real properties of matter.'

The tension appeals to me if only because of its philosophical complexity, apart from my fascination with the character of biolinguistic inquiry as a 'body of doctrine' (Mukherji 2010, Chap. 1). Is there a way, then, of accommodating both the anti-realist and the realist drives of Chomsky in a coherent framework? More specifically, my interest is to resurrect the scientific image from within the sceptical ground charted by Chomsky.

One of the central sceptical points made by Chomsky concerns the notion of intelligibility. As Chomsky puts it, in some fundamental sense the world is unintelligible to us, and 'we have to reduce our sights to search for intelligible theories. We cannot hope to gain comprehension of the world, as Galileo, Descartes and Newton had hoped' (Chomsky 2001). In a way, then, we are compelled to adopt David Hume's position that Newton's discoveries reveal the 'obscurity' in which 'nature's ultimate secrets ever will remain'. This perspective seeks to question what is taken for granted, namely, that 'the natural sciences seek to discover basic truths about the world'. On this realist assumption, the fundamental aspects of the world are progressively unveiled even if 'the scientific enterprise remains open and evolving, and… surprises may lie ahead with unanticipated consequences, as in the past'. The Humean conception of science, in contrast, is that science does not even aim that high. Citing Richard Popkins, Chomsky suggests that 'the secrets of nature, of things-in-themselves, are forever hidden from us'. Thus, we revert to the 'mitigated scepticism' of even pre-Newtonian English science, acknowledging the impossibility of finding 'the first springs of natural motions' (Chomsky 2001). Assuming all this, is there a route from (mere) intelligibility to truth?

3.3 Introducing Schemata

Immanuel Kant studied the issue of intelligibility with much ingenuity. He opens his masterwork *Critique of Pure Reason* (CPR, Kant 1929) with the suggestion that 'our empirical knowledge is made up of what we receive through impressions and of what our own faculty of knowledge (sensible impressions serving merely as the occasion) supplies from itself' (CPR, B2). An immediate consequence of Kant's conception of human knowledge is that, since knowledge is necessarily linked to human receptive and interpretive faculties—*sensibilia* and *understanding*, respectively—the conception of a mind-independent reality, the noumenon, is logically incoherent since, by definition, we cannot describe it (see *this volume*, Chap. 2). In that sense, our conception of reality is restricted to what is intelligible to us, and intelligibility is directly related to our phenomenal grasp of the world. In the absence of a coherent conception of mind-independent reality, on the one hand, and the existence of physics, on the other, a series of questions arise: how is physics possible? What does it mean to say that physics unearths the fundamental principles of the universe if there is no universe to inquire into in the first place?

Kant's interesting answer, as framed early on in his CPR, is that a sense of mind-independent reality is a fundamental human endowment. There is no direct argument, but some hints can be found in his treatment of space and time, two of the fundamental concepts of physics. For Kant, both space and time are *transcendentally ideal*, that is, these are human contributions to the possibility of experience. Space and time are thus conditions of *sensibilia*—a priori *forms of intuition*, in Kant's terminology—where space is the outer sense and time is the inner sense. But they are also *empirically real* in the sense that they foreground the possibility of experience. As he puts it, to account for our empirical experience, we are compelled to assert 'the *empirical reality of space* as regards all possible outer experience' while at the same time we assert 'its *transcendental ideality*—in other words, that it is nothing at all' (CPR, A28). Similarly, 'time is therefore to be regarded as real, not intended as object but as the mode of representation of myself as object. That is, I really have the representation of time and of my determination in it' (CPR, A37). In other words, space and time are viewed as 'real' since that is how we are designed to grasp the world through our experiences, even though they are projections of the human mind.

Extending the framework from space-time to empirical objects, we could say that the transcendental ideality of objects via the human mode of understanding must also project the empirical reality of the object because without this condition we cannot make sense of the content of our experience of objects. In other words, the *intentional* character of our empirical experiences require that there is a world whose objects determine the content of such experiences.

Thus, Kant's general account of physics as a body of empirical knowledge has two parts: constraints on experience and constraints on conceptualization. Roughly, the human mind is endowed with the ability to form strictly universal and, thus, necessary propositions whose empirical content might arise when the concepts occurring in the propositions obey the inner constraints of space and time, the locus of experiences. To put Kant's project in slightly different terms, the concepts employed in scientific propositions obey the overall structure of categories already available to the mind prior to experience; these concepts are filled with empirical content once experiences are harnessed within the manifold of space-time. This was Kant's general expectation.

However, after analyzing the notions of space and time and the general format of categories in which individual concepts find their place, Kant reached the conclusion that the interplay between sensibilia and concepts cannot ultimately be theoretically resolved; the entire construction comes unstuck with what Kant called the 'Problem of Schemata'. Nonetheless, Kant did subsequently identify what he thought was needed to solve the problem of schemata (B 183). He pinned his hope on the notion of *productive imagination*. In a long detour, the notion emerged as follows.

In order for the non-empirical content of concepts to lend structure to the non-conceptual content of experiences, there must be an 'intermediate' level of structures—the *schemata*—that have, *at once*, empirical and conceptual properties. Given the framework within which the *Critique of Pure Reason* was placed, the conception of schemata looked like demanding a resolution of irresolvable opposites: 'no one will say that a category, such as that of causality, can be intuited through sense and is itself contained in appearance' (B 177). In other words, even if the mind is endowed with the category of causality, we don't know how it applies to experience to generate the idea of causal sequences in the world. In despair, Kant opined that

> The schematism of our understanding,… is an art concealed in the depths of the human soul, whose real modes of activity nature is hardly likely ever to allow us to discover, and to have open to our gaze (B 181).

As the post-Kant literature suggested, we may try to illustrate the idea of schemata with the help of, say, visual analogies: a doctor is marked with a stethoscope, an elephant with a trunk, a young girl with short hair, and so on—also known as *exemplars* or *stereotypes*. But we have no idea why these analogies work, assuming that they do at all.

Kant's pessimism about our ability to furnish an account of schemata can be phrased from another direction. As Kant observes (B 179/180), an inquiry into schemata inevitably needs some notion of *productive imagination*, if the schema is to contain the 'conditions of sensibility [that] constitute the universal condition under which alone [a] category [such as number or triangle] can be applied to any object'. This is because, as recognised by philosophers since Plato formulated the one/many problem, appearances, by definition, are particulars and hence their grasp does not by itself contain the resources for reaching *universal* conditions of application. Some additional faculty of mind is required for the necessary step of abstraction (=universalisation). Let us call that faculty *imagination*.

However, by the very nature of the problem, the notion of imagination involved in the formation of concepts needs to be technical, unfamiliar in character. The familiar notion of imagination, Kant points out, is *reproductive* imagination such as when the figure • • • • • gives rise to the image of number five, again a particular: the 'image is a product of the empirical faculty of reproductive imagination' (B 181). It is unclear what 'image' accompanies a 'number in general'. In fact, 'for such a number as a thousand, the image can hardly be surveyed and compared with the concept' (B 180). In general, 'the schema of a pure concept of understanding can never be brought into any image whatsoever' (B 180). Hence, the notion of reproductive imagination is inadequate for giving rise to the schemata. It is hard to miss Kant's refutation of naïve empiricism in these passages.

The notion of imagination involved in schemata, therefore, can only be *productive* imagination in which

> my imagination can delineate the figure of a four-footed animal in a general manner, without limitation to any single determinate figure such as experience, or any possible image that I can represent *in concreto*, actually presents (B 181).

This is what the exemplars mentioned above do, but we don't know how. In other words, on the one hand, reproductive imagination is not enough; on the other, we simply do not know how productive imagination, whatever it is, actually works. In that sense, to repeat, 'the schematism of our understanding,... is an art concealed in the depths of the human soul'.

Nevertheless, it is possible to distinguish between our ability to furnish a 'transcendental' analysis of the concept of schemata, and the *recognition* that, in entertaining empirical knowledge, the human mind has somehow solved the problem of schemata. That is why the exemplars work. To consider a related analogy, there is no doubt that we are able to report on what we see, even if a cognitive account of how we do so remains elusive (Jackendoff 2002). However, notice that, when the problem is solved by the mind, the resources remain within the bounds of intelligibility; at no point does Kant (or Jackendoff) need to postulate properties of the (mind-independent) world itself to show how we come to have knowledge of it. As the philosopher Gottlob Frege suggested over a century later, the mind-internal sense of the object *determine* how the world is like.

As Kant sets up the problem of schemata, two global epistemological moves are barred in effect. For one, we cannot address the problem of schemata by a direct grasp of the categories themselves. Such Platonism is ruled out because categories themselves are just a priori contributions of the mind; they do not apply anywhere until they are filled with the content of sensibilia. For another, we cannot deny that we have at least a restricted universal conception of the world, as the body of Newtonian physics testifies. If the framework of schematism works then we have some hold on how universal conceptions can be reached by beginning with sensibilia alone.

To summarize, the solution of the problem of schemata requires, as we saw, a 'top-down' availability of strict universality plus a 'bottom-up' availability of structured experience: we do not know how the mind puts them together, but we know that it is done somehow. We can think of the laws of physics as the extreme abstract end of this process. Generalizing from categories to laws of physics, thus, the solution requires the incorporation of phenomenological understanding under abstract algebraic representations; in other words, we ultimately need to show how the covering law model of scientific explanation—deriving particular events from universal laws—becomes available.

3.4 Order of Sensations

In an enigmatic and (to my knowledge) largely neglected passage (B 183), Kant makes some preliminary attempts to show how progressive abstractions on sensibilia might give rise to stable conceptions of the world, although, as he insisted, there is no independent hold on these conceptions except through the senses and the a priori contributions of the mind that give rise to these conceptions of the world. As he put it, 'the object which corresponds to sensation is not the transcendental matter of all objects as things in themselves (thinghood, reality)' (B 183). To that end, Kant introduced the notion of 'sensation in general' to suggest that it 'points to being (in time)' (B 183).

Before we develop the suggestion, notice the Heideggerian theme of 'being *in time*'. Kant is suggesting, as noted, that the notion of 'reality' (being) that sensations point to is not being per se; no such thing can be grasped in the framework of the *Critique*. What can be grasped, at most, is some notion of being that is intrinsically related to the 'inner sense'. To put it differently, there ought to be some process wholly in the inner sense (=mind) that points to some object in the outer sense (=external world): the inner sense projects the external world for us. How do we conceptualise this effect while denying the conception of things in themselves?

Within the austere framework of the *Critique*, the only available resource for capturing the required distinction—between the world as projected and the world as such—are the notions of space and time. To recapitulate, both space and time are viewed as a priori *forms* of intuition, that is, both are contributions of the mind for grasping sensible intuitions. In Kant's terms, space and time are *not* properties of things and events as commonly believed; rather, they give forms to sensations as they appear in the otherwise 'blind manifold'. However, for complicated reasons that we need not get into, space is viewed as an outer form of intuition—an 'outer sense'—while time is viewed as an 'inner sense', as noted.

Now, if the postulated sensations in general are to arise from local, particular sensations themselves such as to give rise to some restricted conception of reality in the outer sense, the process that generates sensations in general can only be a function of the inner sense. Since time is the only concept available in the inner sense, sensations in general can only point to being in time. In my understanding, the notion of time plays no other role in the context of schematism beyond providing a 'template'—'merely the form of intuition'—for sensations in general. Hence, I will refrain from examining this (puzzling) aspect of Kant's programme on schemata.

Turning to the basic suggestion of sensations in general, Kant's idea is that sensations come in degrees (magnitudes) ranging from 'nothingness'—empty of magnitude—to complete 'fill out' (Kant's words), even though its representation of the object otherwise remains the same. To use the visual analogy, we can see an object with its full details on close quarters—the details begin to drop out as the object zooms out of view. As the process continues, there comes a point when the last of the details drops out and the object vanishes (ceases to exist in the visual field).

Transferring the analogy to the activities of the inner sense, it looks as though the mind can abstract away from the strong and vivid particularity of sensations—of the *same* object, to emphasize—to sensations that are less and less 'filled out.' Of interest is the point at which the most abstract sensation turns into nothingness (non-being). The *penultimate* state of the inner sense, then, is the most abstract form of being that the mind can construe. Kant's novel suggestion seems to be that this penultimate inner sense of the being can be called 'sensation in general', which presents the object in its maximum generality while continuing to be a sensation.

As just hinted, the most interesting aspect of these suggestions—the reason why I invoked Kant in this chapter—is the apparently counter-intuitive claim that sensation in general—*not* (the original) sensations themselves—points to the being. If I understand the thrust of Kant's proposal correctly, we are led to a rather strong interpretation that the process of abstraction—descent towards non-being—is in fact an indicator of more reality than the full sensation with all the details filled out. In that sense, there is a thin line between being and non-being which is not captured in the thickness of sensations; it is grasped only in their thinness. I will presently cite some textual support for this (strong) interpretation.

This interpretation apparently contrasts with a possible weak interpretation in which the ascent and descent of sensations is merely a way of reinforcing the function of time such that the empirical intuition that sensations can both fill out or reach a vanishing point is a sure indicator of 'something'. Under this interpretation, Kant is leaning on the play of sensations *in time* to thwart the idealist: the idealist has no resources to explain the *growth* (or the *recession*) of the sense of reality even if sensations, at no point thick or thin, signal being.

The weak interpretation is not inconsistent with the strong one. Even if, via the weak interpretation, the up-and-down movement of the content of experiences itself is a pointer to reality, it only follows that grades of sensations point to reality *in a general way*. We do not yet know which grade of sensation points to the 'maximum' of reality. At this point, the strong interpretation suggests that the relation between fullness of sensations and fullness of reality is an inverse one: less sensations, more reality, *up to* the penultimate state, after which the sense of reality disappears. Fullest sensation, then, is not an indicator of being, but of particularity, which is neither being nor non-being—it is just overwhelming appearance. Being, in contrast, is a subterranean conception that does not manifest itself directly in sensations; hence, the need for productive imagination. In my view, this is exactly what Chomsky (1991) meant when he suggested that science is able to extract real properties of matter by abstracting away from the world of ordinary experience.

As Kant puts it, the 'schema of substance is *permanence* of the real in time' (B 183; emphasis added); it is a 'substrate' of empirical determination in time that abides 'while all else changes' (B 183). In other words, the sense of reality (the schema of substance) begins to emerge when 'what is transitory passes away in time', but 'what is non-transitory in its existence' in 'the field of appearance' persists (B 183). Sensations, therefore, contain both transitory and non-transitory aspects; the progressive extraction of the non-transitory elements of sensations by productive imagination yields sensation in general which, as a best fit, is a more compelling pointer to being. The entire weight of the proposal thus hinges on the role played by productive imagination in extracting sensation in general, which is the empirically significant conception of the substratum. Unfortunately, beyond some general comments on causality and the like, Kant does not elaborate on how the suggested abstractions in sensation are to be understood.

To pursue the project, I will suggest that there is a variety of ways in which such abstractions can be generated by productive imagination. Visual images, as noted, are natural grounds for illustrating the desired notion of abstraction. Consider three pictures as follows: (a) a photograph of Mahatma Gandhi's Dandi march which shows the actual Gandhi holding a long stick in his right hand; (b) a photograph of the sculpture by Gautam Pal in front of the Indian Embassy in Washington, DC enacting Gandhi's Dandi march, and (c) a copy of the line drawing of the same marching Gandhi.

It is important to note, pace Kant, that even (a) is a product of imagination with accompanying abstractions: it enables us to 'see' Gandhi without his presence. However, (a) represents *reproductive* imagination; it is a (mere) copy of the original thick sensations. In contrast, (b) and (c) are generated by *productive* imagination in that the mind abstracts away from the original sensations even further while deliberately losing details. In (b), since it is a large sculpture, the mere posture is captured with more 'volume' as the properties of the gait are highlighted by slightly increasing the length of the legs. On the other hand, (c) is just an outline that abstracts away from the volume. In (b) and (c), thus, there is a descent from the original sensation in two different directions. The point is that the descents of sensation bring out more of the 'essence' of the real meaning of (a). In other words, (b) and (c) suggest how the transitory aspects of (a) may be progressively removed by productive imagination to focus on the non-transitory ones.[1]

---

[1]The pictures mentioned in this chapter can be seen either in Mukherji (2009), or by downloading an earlier version of this chapter at http://people.du.ac.in/~nmukherji/work.htm, titled 'Truth, Computation, Intelligibility' (Item 15 under Language and Mind).

Yet, even after the noted artistic abstractions, (b) and (c) continue to be pictures of an individual: Gandhi. Kant's basic problem, however, was to understand how the sensations of an individual—and sensations, by definition, must be of an individual—may give rise to (general) concepts and categories. To that end, it is instructive to look at Pablo Picasso's famous sketches of states of the bull. Picasso's work combines each of the features of abstraction—volume, outline, and expressive gesture—suggested individually in the portraits of Gandhi. Hence, it is a more advanced artistic achievement. These sketches are often characterized as a 'master class' in abstraction:

Bull is a suite of 11 lithographs that have become a master class in how to develop an artwork from the academic to the abstract. In this series of images, all pulled from a single stone, Picasso visually dissects the image of a bull to discover its essential presence through a progressive analysis of its form. Each plate is a successive stage in an investigation to find the absolute 'spirit' of the beast.

In all there are 11 plates. Let me report on a few in Kantian terms. From a younger bull at an early state, Picasso develops a fully-grown bull at the next state to capture the desired expression of aggression: in that sense, this state is an instance of abstract expressionism. As the abstraction proceeds, the representation turns to a 'leaner' bull at subsequent states with much of the transitory features of the adult bull removed. As sensations descend further, the volume gives way to a series of progressively abstract outlines ending in what might be viewed as the 'minimalist' bull at the final state. The suggestion is that, once the last state is reached, we cannot take anything away from it if we wish to retain the bull: any further state turns it into 'non-being'. In that sense, the penultimate state is the being of the bull, its essence.

Picasso's work is a masterpiece because although it does use the natural capacity of productive imagination, it requires skill and reflection of a very high order to actually articulate the sensations in general that the mind grasps. To appreciate this point, consider the picture of a common bull, such as a Texas longhorn, and the well-known Palaeolithic painting of bulls at Lascaux Cave. The painting shows that the human mind was endowed with sophisticated productive imagination from an early stage. Yet the product of that imagination reaches only an intermediate level of abstraction. It needs a Picasso—that is, an entire artistic tradition and a brilliant mind—to take the next few steps.

3.5   Galilean Style

Cedric Boeckx (2006, pp. 96–97) suggested that Picasso's states of the bull throws significant light on how generalizations are achieved in advanced scientific theories. According to Boeckx, Picasso's project suggests an analogy for the sort of abstraction attained from Newtonian mechanics to the theory of relativity, or from the government-binding framework to the minimalist programme in biolinguistics (see *this volume*, Chap. 4). However, despite the genius of Pablo Picasso, the character of abstraction reached so far is not enough for the issue in hand, namely, the notion of scientific truth. Since I described Picasso's method in Kantian terms, the limitation reflects on Kant's programme as well. There are at least two reasons why Kant's project does not apply to scientific inquiry.

First, even if we are able to make some sense, in Kantian terms, of the origin of (general) concepts from the dense particularity of experience, Kant's analysis seems inadequate for the issue in hand, namely, the origin of (advanced) scientific concepts. Kant's method of abstracting sensations in general from particular sensations applies best for what are known as concrete concepts—concepts that cover sensible particulars. That is why Picasso's sketches of the bull illustrated Kant's method so neatly. Postulations of scientific concepts—such as electrons, fields and genes—are not reached by abstracting from particular sensible electrons, fields and genes; there are no such things. These are theoretical entities postulated to explain specific phenomenal effects in a wide range of experiences, often carefully controlled. In contrast, even the most abstract of Picasso's bulls is not a theoretical postulation in the desired sense. Kant's method of productive imagination is not likely, therefore, to illuminate how scientific generalizations are reached by the human mind.

Another related and crucial difference between Picasso's enterprise and scientific theorizing is that science is a body of propositions, not of pictures, although scientific propositions can be aided by or give rise to (abstract) pictorial representations, as we will see. More specifically, in an advanced science such as physics, the propositions are invariably mathematical expressions which are totally devoid of direct pictoriality however abstract. So the sensation in general pointing to being, as captured in a scientific proposition (if at all), is even more abstract than the final, minimalist step of Picasso. As noted, in Picasso's chain of abstractions, when interpreted in Kantian terms, anything beyond that step points to non-being.

How, then, is being captured in science? The answer is obvious from the way we formulated the problem. For physics to represent reality in the abstract, it must focus only on those aspects of sensation in general that are mathematically formulable. In other words, physics so abstracts from thick sensations—the 'world of common experience,' to cite Chomsky (1991, p. 51) again—as to unearth an abstract scheme which some mathematical formulation can generate. Consider the following set of pictures: (a) Photograph of the bronze statue of Lion Pillars at Sarnath with *Asoka cakra* at the base, (b) *Asoka cakra* cropped from the photograph and enlarged, (c) Computer generated image of the *cakra*, and (d) the mathematical equations that generate the form of the *cakra*:

$$x = X + (R - r) * cos`e$$
$$y = Y + (R - r) * sin`e$$

While (a) is an instance of reproductive imagination as with photograph of Gandhi above, (b) represents a part of it focusing on the *cakra* itself. The equations in (d), or variations thereof, generate figures, as in (c), in a computer. My point is that (c) represents, at best, something like the last step of Picasso; hence, it could have been obtained without (d), the equations. It stands to reason that something like (c) must have been entertained by the human mind in order for the scientific mind to come up with a mathematical representation, as in (d), which replicates, to a close approximation, the abstract form already entertained. Yet (c) by itself does not represent scientific progress; it represents at most 'artistic' progress.

In that very specific sense, science takes off from where Picasso leaves sensation in general. We return to the significance of this (scientific) step in the next section. For now, the problem is that the considerations just mooted are necessary, but, by no means, sufficient for physics, because physics is not merely a body of mathematics. The equations in (d) are just mathematical formulae for generating certain geometrical shapes such as in (c); they are not even empirical generalisations. In that sense, these equations do not capture any aspect of reality.

Mathematical physics, at least since Newton—the central topic for Kant—aims much higher. Its mathematical formulae are not only empirical in character, they also signal vast generalizations: Newtonian mechanics, theory of relativity, quantum theory, and now string theory are often viewed as theories of 'everything'. So, the really puzzling feature of the fundamental laws of physics is that they are at once mathematical in character and representations of large aspects of the universe. As we saw in Chap. 2, authors such as Steven Weinberg in fact trace the realistic significance of physics to its mathematical formulations: 'we have all been making abstract mathematical models of the universe to which at least the physicists give a higher degree of reality than they accord the ordinary world of sensations' (Weinberg 1976).

Weinberg and others have called this form of explanation in physics the 'Galilean Style' (Weinberg 1993; Chomsky 1980). The style, according to these authors, works as a foundational methodological principle because of Galileo's insight that nature 'always complies with the easiest and simplest rules;' nature is 'perfect and simple, and creates nothing in vain' (cited in Boeckx 2006, p. 112). How does this (foundational) aspect of physics with its profoundly abstract and, thus, seemingly non-sensual character mesh with the Kantian project? To which being do the constructions of physics point to, and at which descent of sensation? To put it differently, why do the fundamental laws of physics seem intelligible to us, given our bounds of sense?

Once again, the formulation of the problem suggests how it is to be addressed. We recall that in the previous set of pictures of the *chakra* mentioned above, the aspect of 'sensation' in the mathematical symbolism accrued from the fact that these equations formally described a figure that matched, in relevant respects, the descending sensation captured in the (enhanced) reproductive imagination incorporated in (b). There is no 'direct' descent from (b) to (c); if anything, there is, arguably, an ascent from (d) to (c). In that sense, the link between the reproductive imagination and the final productive imagination is, at best, indirect. It stands to reason that, in order for productive imagination to turn even more abstract for approaching Galilean ideals, the link with any possible reproductive imagination is likely to be even weaker, though the link will continue to be non-empty according to Kant's framework.

It seems to me that these austere conditions can be met if we suppose that any further abstraction in the final product requires that the descent of sensations *begins* with *productive* imagination itself, and not with *re*productive imagination. This will contrast with, say, the abstractions captured in Picasso's bull in which the descent begins with *intermediate* states, which in turn are more directly linked to reproductive imagination. The required (ultimate) link with reproductive imagination then will be diffused in abstract thinking; its trace would have been lost in the complex history of human thought. The sensational content of advanced scientific theories in that sense is far more elusive and indirect than the most perspicuous examples of minimalist art.

To illustrate, I will consider the form of representation achieved in Niels Bohr's model of the atom. I have chosen this example because this model is often viewed as a 'planetary' model suggesting that an existing phenomenon—the orbit of planets around the Sun—was the (analogical) trigger for Bohr to develop the idea of electrons circling the nucleus in discrete orbits. The suggestion thus is that Bohr developed his model of the atom by abstracting away from the 'picture' (=reproductive imagination) of the solar system. Without denying that the model of the solar system could have played some analogical role in this case (Hesse 1966), I will suggest that the facts of the case do not support the idea that the scientific thinking *ensued* from some reproductive imagination.

It is easy to realize that there is no direct photograph of the (entire) solar system in the first place for the mind to begin with a reproductive imagination. The solar system is a construction of the human mind acting on pieces of information about planetary motion gathered over several centuries. Any pictorial representation of the solar system is an abstract and elaborate play of human productive imagination that draws the system from known measurements combined with geometrical forms generated from mathematical symbolism.

In any case, for Bohr, even the individual photographs of planets and pictures of partial orbits were not available. Thus, Bohr's conception of the structure of an atom was at best modelled on the analogy of the productive image of the solar system. The 'orbits' in the Bohr model are representations of the excitation and de-excitation of an atom in terms of Bohr's remarkably simple formula, $E = hc/\lambda$, that determines the orbits of electrons around a nucleus. This can be shown in a *graphic*, not a picture, of Bohr excitation states: once a unit of energy is imparted to an electron it occupies a higher orbit around the nucleus; when energy is removed it drops to a lower orbit. This graphical conception gives rise to the 'solar' model.

In Kantian terms, elements of the graphic point to aspects of being—'the real properties of matter'—in an almost total absence of elements of sensation. Yet the point remains that the descending, minimalist sensation captured in the graphic does help generate the artistic conception of the atom which, in turn, is intelligible to us because it relates to the range of sensation that we undergo when we view the model of the solar system. Needless to say, such analogical moves on sensations will be harder to locate for even more abstract formulations of physics.

3.6 Mind and Galilean Style

With the basic elements of the Galilean style in hand, we are in a position to return to the original issue of the truth-bearing character of advanced scientific theories. Recall that we cannot escape the fact that the properties of nature are disclosed to us insofar as they are intelligible in terms of the theories we formulate about them. This restriction to intelligibility is the source of the lie. However, 'intelligibility' is a graded concept. If certain domains of inquiry open themselves for human understanding on the basis of the most stringent standard of intelligibility, then, from within the closed space of intelligibility, so to speak, we get a glimpse of the real. As Kant pointed out, there is no direct inquiry into the real as such; we are bound by the phenomena and the interpretation that we place on them. Yet we hope with some justification that, with the highest standard of intelligibility, the interpreted domain resembles reality as closely as we can get. In effect, as the classical standard of intelligibility of grasping the-world-as-it-is falls, the bar of formal standard of intelligibility is raised a few notches to recover some of the world lost.

The Galilean style offers such a standard of intelligibility. If the phenomena in a certain domain are interpreted with the help of a minimal set of abstract principles that generate, in a long deductive chain, some of the salient features of the phenomena, then the chances are that these principles describe the real properties of nature. This is the best we can get. Post-Galilean physics abounds in such principles; contemporary linguistics is a more recent example. Truth, therefore, is a *consequence* of intelligibility of the highest grade, rather than the mystical property of the mind that grasps the real properties of the world directly. The burden thus shifts to the anti-realist to explain why the adoption of the Galilean style in physics in fact enhances explanatory power. To say that all theories are lies is to miss the grain, apart from suffering from standard charge of inconsistency.

It is interesting that the intelligibility-geared notion of truth just advanced also explains the severe restrictions under which it may be attained. Recall the distinction between Picasso's final, minimalist sketch of the bull and the formal pictorial representation of the *Ashok chakra* generated from a pair of simple mathematical equations. The question is: could we have taken the next, formal, step for Picasso's minimalist bull as well? In other words, is there a mathematical expression that formally generates a figure that representationally matches the artistic conception of the minimalist bull?

Even a cursory look at the last bull suggests that, despite Picasso's minimalist efforts, the form is pretty complex and irregular; the form continues to be 'biological' in character. There could be mathematical expressions that generate this form, but the chances are that they will be highly complex. For the same reason, those expressions will be very specific to this particular form; they are not likely to generalize for the rest of nature. The sinusoidal equations that generate the *chakra*, in contrast, exploit the symmetry and the periodicity of the desired form which replicates in nature in abundance (Stewart 1995, 2001; Carroll 2005; *this volume* Chap. 2). Still, as noted, even the equations for the *chakra* and the form generated thereof are very restrictive in character; elementary geometry and computer science do not pass as theoretical physics.

So, the abstract generalizations captured in theoretical physics need to appeal to much higher orders of symmetry and periodicity. The simplest forms thus are the hardest to identify. Much of the history of science illustrates this point in the actual body of scientific work. Thus Chomsky observes that 'genuine theoretical explanations seem to be restricted to the study of simple systems even in the hard sciences... By the time you get to big molecules, for example, you are mostly describing things'. 'Nature's ultimate secrets', to cite David Hume again, remain shrouded in 'obscurity'. On the rare occasions in which those secrets are revealed, scientific imagination confirms nature's 'drive for the beautiful' (Ernst Haeckel, cited in Chomsky 2001). It took physics hundreds of years of the most imaginative thinking to identify those occasions. How is this perspective likely to apply to the study of mind?

As noted, the contemporary cognitive sciences that broadly represent the scientific study of the mind is a vast array of disciplines, approaches, experimental paradigms and cognitive domains spread across virtually all forms of organisms. This is unsurprising given the immense variety and complexity of cognitive phenomena encountered in the world of ordinary experiences. When we explore the organic world at random without any specific theoretical framework in hand, cognitive effects seem to abound everywhere. Some notion of mind seems to be intuitively at play in the (a) sensory reactions displayed by organisms, (b) complexities of locomotion, (c) identification of food, prey and predator, (d) ability to locate the space for shelter, (e) search for mates, (f) grasping of environmental cues for time of day, season, year, (g) variety of call systems for attracting the attention of conspecifics, (h) rearing of offspring, and the like. The cognitive sciences have thus grown into a bewildering variety of disciplines that address specific domains within the vast canvas.

Following the restrictive Galilean perspective sketched above, it is hardly likely that the present state of the cognitive sciences will eventually furnish a unitary account of the mind from within the approaches currently pursued. To extract a simple and unitary conception of mind from contemporary research, then, some route along Galilean lines need to be charted through the apparent complexity of cognitive phenomena. The first step in that direction is to shift from a general conception of cognitive abilities to a specific conception of the human mind.

As we saw, phenomenally speaking, cognitive abilities abound throughout the evolutionary spread of organisms. No doubt this complex cognitive evolution is a contributory factor to the ultimate emergence of the human mind. Still, there is a clear intuition that something drastic must have happened during the last stages of the hominid evolution to give rise to a form of mentality in humans that is unique in nature. Given severe restrictions on theoretical inquiry, it is unlikely that, in the initial stages of cognitive inquiry, it will be possible to extract some intelligible concept of human mind if we confront cognitive phenomena in full evolutionary complexity. Some principled restrictions are needed in advance to isolate the striking phenomenon of human mentality from the rest of the organic world.

In this context, we recall that language is generally taken to be a unique endowment of the human species. Humans are also distinguished by rich and complex systems of thought and reflective actions. Classical philosophers, especially in the rationalist tradition, drew attention to these (human) 'cognoscitive powers '. The point of interest is that this tradition saw a close connection between these powers. Thus, philosophers like Leibnitz held that language is not only a 'vehicle of thought', but also a 'mirror of the mind'. Another rationalist philosopher Descartes (1637) held that the phenomenon of unbounded and stimulus-free use of 'signs' falls beyond the scope of mechanistic philosophy which otherwise accounts for the 'beasts'; hence, he postulated the novel substance of mind to explain the use of signs which are 'the only marks of thoughts hidden and wrapped up in the body'.

Given the species-specific character of both language and related cognoscitive powers, it is natural to think of these things as intimately related specifically for humans. From this perspective, we may view the development of rich systems of thought as a natural effect as the language system developed through the advancement of the species. By investigating the structure and function of human languages, then, we may be able to form a view of the human mind itself as a defining feature of the species.

The basic clue in that direction is that linguistic theory, that focuses on a central component of the apparently complex human mind, has already met some of the conditions of the Galilean style (Mukherji 2010, Chap. 1). Given the immense complexity of cognitive phenomena and widespread historical scepticism about our ability to study them, it is something of a wonder that the biolinguistic enterprise has been able to extract a strikingly simple picture of human language. This was the source of Chomsky's optimism about the study of language we noted earlier.

Linguistic theory, in its present form of generative grammar, was initiated about sixty years ago. In the beginning, the vast phenomenal complexity of human languages required the postulation of a large number of rules for each of the languages under study (Chomsky 1965). It is generally unknown outside linguistics circles that within a span of a few decades it has been possible to bring down the system to a handful of operations and principles that govern the working of all human languages (Chomsky 1995).

Working under the assumption of virtual conceptual necessity, biolinguistics has been able to characterize the computational system of human language (C_{HL}) which operates under principles which are minimalist in character; in particular, the system consists of the simplest combinatorial operation called *Merge* that functions under a small number of economy principles. With these results in hand, linguistic theory is now in a position to begin to enquire why the language system emerged in the course of organic evolution at all; that is, why the language system has the form it has (Chomsky 2001).

Given the centrality of human language in the structure and functioning of the human mind, as classical philosophers emphasized, and the prospect of attaining some form of Galilean style in the study of language, we might hope to reach some real scientific understanding of at least some aspects of the human mind. However, given the severe restrictions under which the Galilean style works and the intrinsic problems of inquiry on the inner world, much scepticism is in store about what such a theory of mind is likely to accomplish.

References

Boeckx, C. 2006. *Linguistic Minimalism: Origins, Concepts, Methods, and Aims*. Oxford: Oxford University Press.

Carroll, S. 2005. *Endless Forms Most Beautiful*. New York: Norton.

Cartright, N. 1983. *How the Laws of Physics Lie*. Oxford: Oxford University Press.

Chomsky, N. 1965. *Aspects of the Theory of Syntax*. Cambridge: MIT Press.

Chomsky, N. 1980. *Rules and Representations*. Oxford: Basil Blackwell.

Chomsky, N. 1991. Linguistics and cognitive science: Problems and mysteries. In *The Chomskyan Turn*, A. Kasher (Ed.), 26–53. Oxford: Basil Blackwell.

Chomsky, N. 1994. *Language and Thought*. London: Moyer Bell.

Chomsky, N. 1995. *The Minimalist program*. Cambridge: MIT Press.

Chomsky, N. 2000. *The Architecture of Language*. In N. Mukherji, B. Patnaik and R. Agnihotri (Eds.). New Delhi: Oxford University Press.

Chomsky, N. 2001. Language and the rest of the world. Bose Memorial Lecture in Philosophy, Delhi University, November 4.

Dennett, D. C. 1995. *Darwin's Dangerous Idea*. London: Penguin Books.

Descartes, R. 1637/1968. *Discourse on Method* and *The Meditations*. Translated by F. Sutcliffe. Suffolk: Penguin.

Hesse, M. 1966. *Models and Analogies in Science*. Notre Dame, IN: Notre Dame University Press.

Hinzen, W. 2006. *Mind Design and Minimal Syntax*. Oxford: Oxford University Press.

Jackendoff, R. 2002. *Foundations of Language: Brain, Meaning, Grammar, Evolution*. Oxford: Oxford University Press.

Kant, I. 1929. *Critique of Pure Reason*. Translated by N.K. Smith. London: Macmillan.

Mukherji, N. 2009. Truth and intelligibility. In *Science, Literature, Aesthetics*, Volume XV, A. Dev (Ed.). Part 3 of History of Science, Philosophy and Culture in Indian Civilization. New Delhi: Centre for Studies in Civilizations.

Mukherji, N. 2010. *The Primacy of Grammar*. Cambridge: MIT Press.

Osherson, D. 1995. *An Invitation to Cognitive Science*. Cambridge: MIT Press.

Piattelli-Palmarini, M. 1998. Foreword. In *Rhyme and Reason: An Introduction to Minimalist Syntax*, by J. Uriagereka. Cambridge: MIT Press.

Stewart, I. 1995. *Nature's Numbers: Discovering Order and Pattern in the Universe*. London: Weidenfeld and Nicolson.

Stewart, I. 2001. *What Shape is a Snowflake?*. London: Weidenfeld and Nicolson.

Weinberg, S. 1976. The forces of nature. *Bulletin of the American Academy of Arts and Sciences* 29: 13–29.

Weinberg, S. 1993. *Dreams of a Final Theory*. New York: Vintage.

Wilson, R., and F. Keil. 1999. *The MIT Encyclopedia of Cognitive Sciences*. Cambridge: MIT Press.

Chapter 4
Theories and Shifting Domains

> But there is nothing in the real world corresponding to
> language. In fact, it could very well turn out that there is no
> intelligible notion of language.
>
> Noam Chomsky

Suppose we agree that some restricted notion of truth and reality are sometimes available in human theoretical enterprise in terms of the Galilean style. It is still unclear what notion of scientific realism is actually upheld in such theories; how much and which aspects of the world are disclosed in that enterprise?

4.1 Basic Objects

For the non-specialist looking at the scientific enterprise from the outside, the message is that different scientific disciplines carve out different class of objects—or aspects of objects—and events surrounding them in the world; a specific discipline gives an account of this part of the world in terms of the theories of the discipline. Thus, physics describes the physical aspects, chemistry the chemical aspects, linguistics describes human language, and so on. Since these disciplines carve out constituents of the world, we may identify the disciplines in terms of what they carve out. For the purpose of exposition, let us call them *basic objects*.

Philosophers sometimes try to downplay the idea of disciplines. For example, Willard Quine once remarked that disciplines are for the deans. Nonetheless, as we will see, the notion of a discipline is non-trivial; it plays explanatory role in a variety of human inquiry. We discuss them as occasions arise. For a quick glimpse of what may be at issue, consider the notion of interdisciplinarity. It is obvious to me that interdisciplinary studies, like interracial marriages, presuppose prior existence of disciplines (and races).

This is a revised version of a paper published as Mukherji (2001).

© Springer Nature Singapore Pte Ltd 2017
N. Mukherji, *Reflections on Human Inquiry*,
DOI 10.1007/978-981-10-5364-1_4

Apparently, the idea of basic objects marking a discipline is ambiguously related to another closely similar idea. The literature on scientific realism sometimes suggests that some strong version of realism can still be upheld if scientific theories are viewed as unearthing some fundamental aspects of the world in terms of the entities postulated by a theory: *entity realism*. Since the thesis comports with common sense, I will refrain from citing literature that promotes it. The point, albeit vague, is: what is the connection between theory-entities and basic objects of disciplines? The query seems significant since, if we *identify* basic objects with theory-entities, basic objects—and hence the discipline—changes as soon as theories *of* a discipline changes. Notwithstanding a variety of nudges in that direction from the Kuhnian literature, the suggested identification is uncomfortable because it casts a doubt on our commonsensical idea of the world: the world is what it is not because Paul Dirac made a comment about it.

Despite the suggested distinction between theories and disciplines, the starting point can only be the content of theories because that is all that we have in hand; basic objects of a discipline can only occur in theories *in* the discipline. So, is there a way of reaching the idea of basic objects from the content of theories? To refresh, the idea of basic object(s) of a theory is seductive in at least two ways. First, if a theory has basic objects which *individuate* the theory in telling us what the theory is about, then, if there is something like a universal theory of everything that there is, then the global theory will capture the fundamental furniture of the universe. In saying this, I am leaning on a conception of *universal* theories; examples include exactly Newtonian mechanics and relativity theory. Einstein suggested that quantum theory is not universal in this sense; it is an incomplete theory. So, Einstein must have had a prior conception of the physical (aspects of) the universe such that he was convinced that quantum physics, *with its entities and events*, is not (properly) describing the universe. I am *not* thinking of *unified* theories whose existence is in dispute in any case. Even for less ambitious scientific theories, the basic objects, if any, will sketch a local picture such as what the organic world looks like, what cognitive systems look like, what human language looks like, etc. And in each case, a practitioner within a certain discipline may dispute whether a construct proposed by another practitioner belongs to that disciplinary realm.

Second, the conception of basic objects allows a robust realist interpretation of theory change. If theories have basic objects that individuate a theory in some fundamental way, then we can try to keep the effects of much of the empirical features of theory-change to its *non*-basic ones. Keeping track of the basic objects across theory-change thus gives us a handle on discipline-identity—quantum physics, polymer chemistry, biolinguistics, etc.—in terms of their domains, that is, what they are about. However, in order to force a basic/non-basic distinction in the objects of a theory, we need first to convince ourselves that theories in fact *have* objects; they are not mere instruments that enable systematic computation over a choice-set of symbols invented by someone. Once we are filled with conviction, we could try to split this class into basic and non-basic subclasses.

For example, the contemporary discipline of generative grammar asserts that recursion is a basic property of language. As Chomsky (2015, p. 4) formulates it,

Each language provides an unbounded array of hierarchically structured expressions that receive interpretations at two interfaces, sensorimotor for externalization and conceptual-intentional for mental processes.

There are a range of theories in generative grammar—Standard Theory, Revised Standard Theory, Revised Extended Standard Theory, Government-Binding Theory, Bare Phrase Structure Theory, Phase Theory—within the 'Chomsky' side of the discipline alone. Each theory represents the basic property of recursion somehow within its vocabulary. This constraint on grammatical theory is so severe that if a theoretical move attempts to deny that recursion is a property of language, the move is seen as falling outside the discipline. Thus, Everett (2005) suggested that the Amazonian language pirāhā does not have recursion. The response from Chomsky and his followers was that, even if Everett's data are fine, it is of no consequence since recursion is a basic property of the faculty of language, not of speech-events in the world. In other words, recursion marks the *domain* of a specific aspect of language, in the sense of *domain* as follows.

4.2 Disciplinary Domains

It may be instructive to begin with a fairly commonsensical view of scientific theories aired, say, by the philosopher Jerry Fodor (1994, p. 3).

Empirical explanation is typically a matter of subsuming events (states, etc.) in the domain of a science under laws that are articulated in its proprietary theoretical vocabulary.

The view is commonsensical in that it does not invoke a technical notion of theories as sets of sentences. The view concerns directly what theories are, not how they look like. What are the basic objects of a theory in this conception of theories?

The central notion in Fodor's conception of theories—namely, empirical explanations via laws—is that a theory is geared to a *domain*. The conception of a domain here is a hard-headed one having to do with domains of experience, domains of reality, and the like. I will settle for domains of reality because events are metaphysical entities, not epistemic ones. Thus, what contains them ought to be metaphysically construed as well. So, there are domains of reality out there, whether we experience events in such domains or not.

This realistic conception of domains gives us an immediate handle on the issue of individuation of theories. Theories are now individuated in terms of the events that are explained by the theory via its 'proprietary theoretical vocabulary'; the proprietary vocabulary becomes primitive vocabulary, if the theory has been formalized. This could only be if the said events could be *described* by the vocabulary of the theory. This, in turn, could only be if the vocabulary of the theory consists in part of terms that pick out certain objects in the domain such that interactions between these objects in time give rise to the events that ultimately fill the domain. A theory can now well be viewed as true *of* such events, *simpliciter*. These objects could be bodies in relation with one another, or bodies and forces, or just forces acting on each other, or perhaps they are other things like mental particulars; it need not matter. All that is required is that the objects perform the role of values in a true theory, assuming that the theory has been formalized (Quine 1953).

We thus get a natural fit between a domain and the vocabulary of a theory. To use a familiar metaphor, we think of domains as areas carved out from the rest of reality. Domains in that sense have a geometry; we think of objects as marking out this geometry like pillars on a field. No doubt, various kinds of objects can do the job—bushes, for example. So, it is hard to think of a *given* set of objects as the only one that does the job. Yet the very geometry of the domain will impose rather severe constraints on what these objects could be. It is possible, then, to extract from these constraints something like the very condition of objecthood in that domain. Each theory claiming to focus on the domain will describe the objects in the domain in terms of its vocabulary. So, the description of the objects may differ in a variety of ways, but the domain remains fixed.

These objects will show up in *any* theory that explains events in a given domain. We thus have exactly the sort of distinction between basic and non-basic objects of a theory that explains theory-change without falling into (metaphysical) relativism. Notice that the preceding line of thinking does not give us any independent hold on *theory*-identity. As Fodor correctly noted, the notion of a domain attaches to a given *science*, not really to the theories within that science. So, the most we have got is some conception of basic objects of a given science. With this I have given up the idea of a basic object of a theory and have settled for basic objects of a science although, no doubt, the basic objects will be *expressed* by the primitive vocabulary of a given theory at a time, perhaps as values of its variables; how else?

Not every joint of the preceding picture is exactly clear. But I stated them anyway because they are widely held in a pretheoretical way. We seem to need fairly secured disciplinary notions of physics, chemistry, botany, neuropsychology, linguistics, and the like to raise questions about their individual origins, interdisciplinary possibilities, reduction, autonomy, unification, and much else. For example, when asked about the possible event of nuclear winter and the extinction of dinosaurs due to the crash of a meteor, the physicist-writer Carl Sagan replied that the physics in the two cases is the same. For Sagan, then, certain events involving very different objects such as nuclear devices and meteors fall under the unitary study of the physical. But in the pedagogical form, disciplines are also marked by their theories, which keep changing drastically. So, it is natural to seek some separation between disciplines and their theories in terms of some pretheoretical stretch of the world that each theory in a discipline attempts to describe.

Suppose we grant all this so that we can proceed. Let us not get bogged down with questions about what these basic objects of a science are, how much degree of freedom they allow in the construction of theories, whether these are always abstract objects as suggested, and the like. These questions need not detain us because from here on I am going to focus on the notion of a domain itself. In any case, I hope the following example makes it all clear.

4.3 From G-B to Minimalism

The picture just sketched assumes, *à la* Fodor and common sense, that a science, now viewed diachronically as a cluster of theories, is individuated in terms of a domain: when domains differ, sciences differ; also, when domains overlap, sciences overlap. Is that the correct picture of an ongoing science? In this section of the chapter, I will examine the question just asked with respect to some recent developments in linguistic theory. That is, I am going to see if the bit of historical data I am going to look at fits with the idea that a science can be individuated in terms of its domain.

In the late 1970s, Noam Chomsky proposed a theory in which the concepts of government and binding played a central role. So, in popular parlance, as well as in print, the theory came to be known as *Government-Binding Theory* (Chomsky 1981, 1982). I will presently make an informal sketch of these concepts. However, before we do that, it is important to note the radical nature of these proposals with respect to traditional conception of linguistics. At a number of places, Chomsky has claimed that the theoretical framework in which the concepts of government and binding play a central role—called the *Principles and Parameters framework*—is a radical departure from 'thousands of years' of research on language (Chomsky 1991, p. 23). The framework departs from traditional concerns, such as the ancient Paninian grammar, in at least three significant and interrelated ways:

(a) The framework concerns *knowledge* of language, that is, states in the mind of native speakers, rather than language as an external object with properties of sound and meaning.
(b) The framework attempts to identify the genetic properties of the species rather than the properties of individuals and communities.
(c) The framework views notions such as Oriya, English and Sanskrit as non-theoretical and on a par with such notions as 'large molecules' and 'terrestrial animals'.

Therefore, on Fodorian grounds, current linguistic theory differs totally from traditional ones in that the domains differ sharply. The events studied by traditional grammar were speech-events taking place in the world; the events that interest Chomsky happen inside the mind of the child. So, there is not even a partial overlap. How, then, do we understand Chomsky's statements such as the study of language is an 'ancient' one that 'goes back thousands of years?' In what sense are Panini and Chomsky joining hands in a common enterprise?

One could reply, plausibly in my view, that issues about the basic objects of a theory meaningfully arise only for formal theories in which its primitive vocabulary is explicitly identified. Thus, it is unfair to raise this issue *across* a formal framework and others that are largely informal commonsensical approaches on vaguely defined phenomena. In this view then we should restrict ourselves only to the Chomskyan framework and examine *its* basic objects, if any. I doubt if any empirical theory is strictly formal in the sense just demanded, even if we ignore lessons from Gödel. Further, I doubt whether die-hard Paninians will agree that Panini's was an informal commonsensical approach. Let us set these doubts aside and proceed.

Returning to the theory of government and binding, the concept of government was introduced as follows. In a grammatical explanation geared, say, to the structure of a syntactic tree, various conceptions are needed to relate syntactic objects at the nodes of the tree. Suppose we wish to represent the syntactic structure of a simple clause *that John liked Mary*. Here *that* is a complementizer (COMP), *John* and *Mary* are nominal expressions (NP), *like* is a verb (V), and *-d* is a verbal inflection marking past tense (INFL). These syntactic objects may be used to construct a tree for the mentioned clause. See Figure. Once the tree is formed with the noted hierarchy displaying a variety of projections, some structural relations between the syntactic objects may now be postulated to capture various grammatical facts.

The central notion is that of c-command which relates syntactic objects, roughly, under maximal projection. C-command will thus relate fairly 'distant' objects, say, objects at specifier (such as INFL, COMP) and head (such as V, N) nodes of a given phrase.

Basic Syntactic Tree

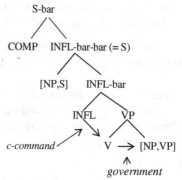

However, it was observed that important syntactic generalizations could be reached much more 'locally', that is, within 'flatter' parts of a tree. The concept of government is designed to capture such local relations, in part, by way of, roughly, mutual c-command, and other things. For example, the relation of government obtains between a verbal head V and its complement, say, an NP (*see* [*the man*]). For two mutually exclusive elements α and β, let us say 'α c-commands β' just in case every maximal projection dominating α dominates β. We say that an element 'α governs an element β' just in case α and β c-command each other and α is a head, that is, $\alpha \in \{N, V, A, P, INFL\}$ which means that α is one of the basic syntactic categories: noun, verb, adjective, preposition, and tense. Thus, in the figure, INFL c-commands V, V does not c-command either INFL or [NP, S] though V c-commands [NP, VP] which also c-commands V. Therefore, INFL does not govern V but V governs [NP, VP]; [NP, VP] does not govern V since the former is not a head.

However, another special sort of relation is needed to capture generalizations regarding the distribution of noun phrases (NPs) themselves, especially when they get co-indexed for a variety of reasons. Pronouns as in *John$_i$ thought that he$_i$* [John] *needed a shave* and reflexive pronouns as in *John$_i$ decided to shave himself$_i$* are paradigmatic examples of such dependent elements as co-indexing shows. Notice that the pronoun *he* may have a disjoint reference as in *John thought that he* [Bill] *needed a shave*, but the reflexive anaphor *himself* can only refer to John; in contrast, in *John decided to shave him*, *John* and *him* cannot be co-indexed, *him* has to be someone other than John. The relation of government, though local, is too general to capture just such information. Also, the generalizations required here need not obtain in local domains; *the children$_i$ thought that pictures of each other$_i$ were on sale*. Additional constraints on c-command give the relation of binding between co-indexed NPs. Thus, both the relations of government and binding are required in the system.

The concepts of government and binding were thought to be so central to the linguistic enterprise that, in a famous lecture, Chomsky (1991) admonished his followers in the following words which I cite at length:

> Such terms as 'government-binding theory' should be abandoned... Insofar as the concept of government enters into the structure of human language, every approach will have a theory of government... Similarly, no approach to language will fail to incorporate some version of binding theory, insofar as referential dependence is a real phenomenon to be captured in the study of language, this being a common enterprise. There are real questions about government and binding, but no tentative set of hypotheses about language has any proprietary claim to these topics.

So Chomsky is suggesting that the concepts under discussion 'enter into the structure of human language' , a study of which is a 'common enterprise'. These concepts relate to 'real questions' which 'no approach to language will fail to incorporate'. In our terms, these concepts thus signal the basic objects of the linguistic enterprise itself, not just of specific theories—but of 'tentative set of hypotheses'. They are needed to capture the events in the domain of language to which every theory of language from hereon will be geared. In sum, we get everything that we wanted of basic objects and their ability to supply disciplinary identity. This was in a lecture delivered in 1989, which was subsequently published in 1991.

In 1992, that is, within three years, Chomsky circulated a paper titled 'A Minimalist Program for Linguistic Theory'. This paper was subsequently included as Chap. 3 in Chomsky (1995). In this book, he proposed another ground-breaking framework in which, as he summarized later (Chomsky 2000, p. 37),

> there's no government, no proper government, no binding theory internal to language, and no interactions of other kinds.

The theoretical reasons why the minimalist programme (MP) does not have government, and why binding theory has been taken away from language and has been placed elsewhere, are far too involved and technical for a quick exposition here.

Roughly, the basic idea is that these concepts are no longer required since: (a) the programme dispenses with the notion of grammaticality which required binding theory; and (b) the programme dispenses with the levels of representation, such as d- and s-structures, where the concept of government played a crucial role. In some global sense, the theory still explains the 'phenomena' covered earlier; in fact, the claim is that it covers much more. But that notion of phenomena can no longer be captured in terms of the proprietary vocabulary of government-binding theory. In short, the *domain* has shifted radically from the G-B framework to the minimalist one. What identifies the discipline of linguistics then?

4.4 Objections and Replies

Several objections need to be considered at this point. First, one could generally object that the sketch of the minimalist programme does not show that the *phenomena* of government and binding have disappeared from view. One could still have government if one wanted to, but it is no longer required since better and more economical devices to capture the *same* phenomena have been found. There are several problems with this objection. Linguistic behaviour of people is certainly the 'phenomena' that linguistic theory, in any version, tries to explain. But that, as in any science, is just a starting point since linguistic behaviour is the data. What we are interested in are the basic objects of *theories*, not of 'experiences'. Which objects make a certain stretch of experience possible is exactly what a theory tries to explain. It is obvious that if experiences came marked with their objects, no science would have ever been needed.

Second and more specifically, one could say that the historical sketch goes to show only that the chosen concepts are not basic, not that there are no basic objects. For the concept of c-command which was central in defining both government and binding is still available in the minimalist programme. So c-command is one of the basic objects (rather, relations); government and binding are not. This objection, in recent philosophical parlance, simply kicks the problem upstairs, for the 'basicness' of government and binding has been passed on to the 'basicness' of c-command.

The objection assumes that the concept of c-command remains invariant across G-B and Minimalism such that we could get back government and binding if we wanted to. This is far from the case, however. First, there is no doubt that a relation which is continued to be called *c-command* is available in MP. Let us also grant, questionably, that the empirical effect of MP c-command is equivalent to GB c-command. Yet the MP c-command is *defined* in MP terms—that is, in terms of targets and visibility in structures formed out of merging two earlier structures—which are not available in GB. So if we define government with MP c-command, it is unclear if we are describing the same phenomenon. Further, MP works with a primitive notion of locality which is significantly different from government locality. Thus, there is no natural way of getting one from the other, and, as suggested, the exercise is not even required since the derived notion of government, if any, does not have a function in MP.

The situation with binding is somewhat different. As suggested, binding theory no longer constrains operations and representations in the computational system. So, even if it is available, it is no longer a basic relation in linguistic theory. In fact, it is not even available despite the availability of MP c-command. Recall that binding theory requires two clauses: c-command and co-indexing. MP does not have indexing in the first place for framework internal (=minimalist) reasons. Some version of binding theory applies to the outputs of the computational system and, hence, indexing needs to be introduced in some way. There are several options available here, however; say, linking, or referential dependence, which are very different from GB indexing. These notions in any case are not available as basic notions in MP. In my view, similar remarks apply to the status of the alleged basic property of recursion as it is reformulated repeatedly through changing theories, from Standard Theory to the minimalist programme. But the topic is too involved for a quick discussion here (Mukherji 2003, Arsenijevic' and Hinzen 2012).

A third and more potent objection could be that, despite internal shifts in the vocabulary, the *domain*, in some global sense, remains intact. Aren't all frameworks in linguistics geared to the domain of *language*? If they are, then there must be basic objects which individuate the domain of language. Can we, for example, give up concepts like nouns, verbs, prepositions, reflexives, etc. while continuing to do linguistics? A little later, I will address the issue of whether linguistic theory, notwithstanding what it is *called*, is necessarily geared to the domain of language. For the moment, let us grant it. Even then the objection amounts to a stipulation.

The domain of language is a pretheoretical conception, just as the domain of physical theory is pretheoretical. Some array of experiences, expectations, etc. no doubt give rise to such conceptions. The task of a science is to *interpret* and examine them to see whether they are valid. If their validity is taken for granted, then the issue of basic objects ceases to be an empirical issue. Basic objects, if anything, are projections of *theories* suitably formalized, *not* of expectations. If that was the case, then the growth of plants would have continued to fall within the domain of physics, as Aristotle thought, since motion—viewed as displacement over time—is involved there. Similarly, we saw that the advent of GB signalled sharp change in the domain itself, namely, in the very notion of language. So, ultimately, we want physics and linguistics to tell us what their basic objects are, and the sense in which they are 'basic'. Domains of science are constructions, not given in advance. The interesting question under discussion is whether such constructions reach a stable core.

Turning to the more specific thrusts of the current objection, it is not even obvious that the conception of a domain for linguistics forces a basic vocabulary such as nouns, verbs and reflexives. Returning once more to the beginnings of the current linguistic enterprise, the basic reason for advancing transformational grammars against phrase structure grammars was that the latter did not give natural explanations of linguistic facts such as passive constructions. Roughly, the passive structure *Mary was kissed by John* ought to have deep structural connections with the active *John kissed Mary*. In cognitive science this came to be known as the 'systematicity requirement' (Fodor 1998). Clearly, one could show the structural link between these constructions by isolating their *units* of construction, rearranging them, and mapping one sequence of them to the other. The syntactic categories, with which these mapping functions (=grammatical transformations) were defined, happened to be things like NPs, VPs, and a host of others. NPs and VPs themselves were built out of smaller, atomic units such as n(oun), v(erb) etc. So, it might well seem that these last-named are the basic objects of the domain of language itself, rather than of a given theory.

I took pains to discuss the role of these basic linguistic concepts in early generative grammar to bring out the (by now) familiar point that, ultimately, the identity of a concept is to be understood in terms of the theoretical role it plays. So, as in the case of c-command above, if the role varies sufficiently in a succeeding theory, then doubts arise as to the identity of the concept *across* the theories. Now, according to Chomsky, the central difference between the Principles and Parameters framework and its ancestors is that the former dispenses with the very idea of construction-specific rules: notions such as active, passive, reflexive etc. are now treated as 'taxonomic artifacts' on a par with things such as pet fish and shady tree.

The more recent systems allow only one transformational rule—Move-alpha or Affect-alpha, currently viewed as Internal Merge—that is not sensitive to the *type* of syntactic category it works upon. Therefore, the classical labels such as *noun* and *verb* do not play any theoretical role in either defining constructions or transformations. Thus, the chances are these are very different basic objects, if at all. No wonder then that the concept of a phrase itself differs markedly between classical transformational grammar and GB, and between GB and MP. The topic is too technical for informal exposition here. But the basic point should be clear already.

So far, we have granted that all theories of language focus upon a pretheoretical domain called *language* even if, as we just saw, we are unable to outline the geometry of this domain with a 'proprietary' set of basic objects. The notion of language thus lacks stable theoretical content. Are we compelled to hold on to this pretheoretical notion, whatever it is? No doubt we *start* with some such notion. But as Larson and Segal (1995, p. 8) point out,

> In the process of constructing a rigorous and explicit theory, we must be prepared for elements in the pretheoretical domain to be reanalyzed and redescribed in various ways.

This much is almost obvious. What is not so obvious lies buried in historical facts about the growth of a theoretical enterprise. It often happens in the historical process of 'reanalyzing and redescribing' an initial domain that 'data' earlier thought to be fundamental turns out to be invalid or irrelevant on closer theoretical scrutiny.

More significantly, during the same process, new data begin to emerge that were not even visible earlier. This requires proposals for new theoretical tools, and the entire theoretical machinery needs to be redesigned to accommodate these new facts coherently. Thus, as the phenomenal field changes, the domain shifts gradually. As Chomsky once remarked, the field changes the moment a graduate student enters your office. We have seen several examples of such domain shift. The logical consequence of these incremental shifts is that there comes a point when so much of the earlier domain has shifted out of view, and so much of a new one has got into focus, that the unity of the changed theory can only be understood in terms of a fresh domain, leading to a new discipline. The exclusion of growth of plants, and inclusion of planetary motion in the domain of physical explanation, led to the separation between biology and physics, as we knew them until this century. By parity of reason, then, there is no more basis for thinking of *current* domains of physics and biology as absolute. There are many examples in the history of science which exemplify the point. In fact, without such a dynamic, it is difficult to understand the proliferation of disciplines and subdisciplines, as we find them today.

With this general historical scenario in mind, let me turn briefly for the last time to some features of the current minimalist programme to address the issue of the domain for linguistics. We saw that no interesting sense can be given to the notion of proprietary vocabulary to identify the domain called *language*, although it is beyond dispute that linguists study things like English, Oriya and Japanese for whatever theoretical goals they have in mind. Yet, as the enterprise progresses, *aspects* of these pretheoretical entities, that are brought under the scope of a theory, keep changing.

Thus, for linguists working within the principles and parameters framework, *Oriya* means a combination of parametric values instantiated in the mind of a child. In MP, these values are located essentially in a small discernible part of the lexicon. The rest of the system, namely, the computational system, is entirely universal and is immune to the differences between pretheoretical objects. Arguably, then, the computational system, whose architecture is the central focus of MP, could very well apply to objects outside the *entire* pretheoretical set. Suppose it applies to some aspects of the domain of music (Mukherji 2000, 2010). What is the domain of this new theory? The only legitimate answer is that, it is *neither language nor music* when conceived pretheoretically. Theoretically, it is just a computational system in nature, probably restricted to humans.

Even when we restrict attention exclusively to 'language' without extending the reach of the evolving theory to music, etc., by now it is unclear if any intuitive notion of language survives in the theory anymore. Suppose we start the study of language with the common notion of language—approved apparently by Aristotle—that language is a system of sound-meaning correlations. As we will see in some detail in the next chapter (Chap. 5), the notion of meaning captured in linguistic theory falls far short of the common expectations around this notion.

In linguistic theory, meaning is an extremely abstract notion that has nothing to do with objects in the world, mental images, people's beliefs and intentions, social norms, and the like. The 'meaning'-part of the sound–meaning correlations is now captured in LF representations, representation of Logical Form, which can be viewed as an extension of syntax. The operative notion under study, then, is a restricted notion of *grammar*, rather than some thick notion of 'language.' In fact, Chomsky suggests that only grammars are likely to have a real existence; the notion of language may not even be intelligible. Grammars then establish correlations between sound and a very restricted unfamiliar, theory-guided notion of meaning. But even this picture seems to be under drastic revision in more recent work.

Without going into technical details, let us simply record that, throughout the history of contemporary linguistics, it has been observed that the phonological part of the grammatical computation may be radically different from the rest of the computational principles (Bromberger and Halle 1991; Halle 1995; Harley and Noyer 1999). Moreover, representation of sound is linear in its temporal form while grammatical representation obeys only hierarchy. As Berwick and Chomsky (2016, p. 8) point out, the sentence *birds that fly instinctively swim* is ambiguous: birds either fly instinctively or they swim instinctively. However, the sentence *instinctively birds that fly swim* can only mean birds swim instinctively, even though *instinctively* is more distant from *swim* than *fly*. This is because, if we analyze these sentences in terms of their grammatical structure, it turns out that in the second case *instinctively* is actually closer to *swim* than *fly* (Berwick and Chomsky 2016, p. 117). There is some neurological evidence—always controversial—suggesting that grammatical/hierarchical processing is largely independent of processing of sound.

These concerns have led to the idea that the sound system of human language is ancillary to the grammatical system. The grammatical system is basically designed for construction of (internal) thought. The sound system is something like an evolutionary afterthought geared to articulation or externalization of thought. If humans were endowed with some mechanism for telepathy, the sound system, or other systems of articulation such as gestures, would not have been needed. So much for Aristotle's idea that language is a system of sound–meaning correlations.

I personally do not think that sound is ancillary to the language system; my view, for what it is worth, is that the computational system of language could not have evolved without the sound system evolving in tandem. But we are not engaged here with the correctness of specific proposals in linguistics. We are concerned with the state of the art of the discipline of linguistics. Since the classical notion of language is no longer in use in the current state of the discipline, doubts arise as to what the discipline is currently talking about, if anything.

4.5   A Final Objection

There is a fourth objection to the historical example discussed in this chapter. It needs to be treated separately because the objection has to do not with some aspects of the example, but with the validity of the example itself. One could argue that a meaningful discussion of the basic objects of a scientific theory ought to focus on matured and hard sciences, not on theoretical enterprises in their infancy. A beginning science is naturally unstable, and it has to go through several 'upheavals' before it settles down to a coherent picture of the reality it projects. This response is interesting because it takes seriously the idea of a science delinking itself from pretheoretical conceptions. Recall that much of the preceding discussion was also based on this idea. So, here the point of the objection is that until some time has passed to allow a science to find its own nest, so to speak, we cannot legitimately talk about its basic objects. The short history of formal linguistics violates this condition.

Again, there are several difficulties here. First, delinking a science from pretheory only robs the science of a domain it can cling on to; the delinking *does not* entail that the science finds its own *sustainable* domain. Quite the contrary, as we saw. Since we failed to find any stable notion of a domain within the course of the formal theoretical enterprise, we appealed to the pretheoretical conception as a last resort, to allow the greatest possible room for the issue to maneuver. Having failed in that move, we could conclusively reject any coherent notion of a stable domain. Delinking, thus, is not a move towards stability; just the opposite, in fact.

Moreover, granting that we can form some conception of suitable 'time' to have elapsed before we talk about basic objects, it is not clear at all what contribution does the passage of time make on this issue. In fact, one could argue, plausibly in my view, that the study of basic objects ought to be focused, if at all, to the *early* phases of formal theorizing for the best results. It is well known that metaphysical battles are fought in science only at the early phases, and at a very late phase when faced with a crisis. Since the latter phases in fact cast doubt on the very availability of basic objects, it follows that the search could be successfully conducted only in the early phases. In the beginning, theories tend to be innocent, self-critical and open to radical reformulation. The events of their tortuous delinking from pretheory are still fresh in the collective memory of the enterprise. The theory is still simple enough for us to look at its foundations with sufficient accuracy and coverage. No wonder biologists prefer baby mice. The trouble is that it is hard to find a case where an enterprise is sufficiently formalized *and* is in its infancy. Contemporary linguistics provides that rare opportunity.

As the science matures over centuries and gets 'harder', the theoretical edifice becomes enormously complex with many layers and practices filling the space. Some of these acquire enough autonomy of research to resist almost any 'outsider' attempt to look at its foundations; the 'insiders' had by then developed enough vested interests in the protection of the sector they occupy. A mature science, then, is a breeding ground for dogma. Centuries of success give rise to the illusion that the science has finally hit the 'truth', that its basic objects are inviolable. The intimate details of subtle shifts of research strategies that lead to shifting domains are lost from the memory of the enterprise, or, they become too scattered for a collective view. Thus, when a crisis hits the enterprise, it gets difficult to retrieve the track record from the archives.

Finally, it is not even clear that the central features of the picture that we sketched for the linguistic enterprise do not apply to the more advanced sciences. For a quick example, consider some of the recent remarks by Roger Penrose (2001). Discussing the measurement problem in quantum theory, and the prospects for its unification with the theory of relativity, Penrose suggested aiming for a 'new physics'. In this new physics, quantum theory and, say, gravitation will be properly unified, that is, we will expect gravitational effects at the quantum scale. For that to happen though, the current scales of both quantum theory and relativity theory are insufficient. If we take a free electron (current scale of quantum theory), we get the relevant quantum effects, but the gravitational effects are too small. If we take a cat (current scale of relativity), quantum theory produces paradoxes. So, we settle for an intermediate scale, say, the scale of 'a speck of dust'.

With a speck of dust, Penrose suggests, you can start to ask the question, 'Could a speck of dust be in this place and in that place at the same time?' 'My arguments', he continues, 'would say that, at a certain level, you will start to see differences from the quantum procedure, and, at this level, you can actually compute on the basis that this is a *gravitational effect*, that somehow it is part of this union between, on the one hand quantum mechanics and, on the other hand, Einstein's general relativity.'

The point to note here is the familiar one: a speck of dust is not in the domain either of quantum theory or relativity theory, as these are currently conceived. The effect sharpens if we include, as Penrose suggests elsewhere (1994), *consciousness* to fall within the scope of new physics. If Penrose is right, we ought to retrain our minds, without falling into Buddhism, to conceive of a domain that has consciousness as well as a tiny speck of dust. That ought to take some shifting in our conception of physics.

References

Arsenijevic´, B., and W. Hinzen. 2012. On the Absence of X-within-X Recursion in Human Grammar. *Linguistic Inquiry*, Volume 43, Number 3, 423–440.

Berwick, R., and N. Chomsky. 2016. *Why Only Us: Language and Evolution*. Cambridge: MIT Press.

Bromberger, S., and M. Halle. 1991. Why phonology is different. In *The Chomskyan Turn*, A. Kasher (Ed.), 56–77. Oxford: Basil Blackwell.

Chomsky, N. 1981. *Lectures on Government and Binding*. Dordrecht: Foris.

Chomsky, N. 1982. *Some Concepts and Consequences of the Theory of Government and Binding*. Cambridge: MIT Press.

Chomsky, N. 1991. Linguistics and adjacent fields: A personal view. In *The Chomskyan Turn*, A. Kasher (Ed.), 1–25. Oxford: Basil Blackwell.

Chomsky, N. 1995. *The Minimalist Program*. Cambridge: MIT Press.

Chomsky, N. 2000. *The Architecture of Language*. In N. Mukherji, B.N. Patnaik, R. Agnihotri (Eds.). New Delhi: Oxford University Press.

Chomsky, N. 2015. *What Kind of Creatures are We?*. New York: Columbia University Press.

Everett, D. 2005. Cultural constraints on grammar and cognition in Pirahã. *Current Anthropology* 46: 621–646.

Fodor, J. 1994. *The Elm and the Expert*. Cambridge: MIT Press.

Fodor, J. 1998. *Concepts: Where Cognitive Science Went Wrong*. Oxford: Clarendon Press.

Halle, M. 1995. Feature geometry and feature spreading. *Linguistic Inquiry* 26: 1–46.

Harley, H., and R. Noyer. 1999. State-of-the-article: Distributed morphology. *GLOT International* 4 (4): 3–9.

Larson, R., and G. Segal. 1995. *Knowledge of Meaning*. Cambridge: MIT Press.

Mukherji, N. 2000. *The Cartesian Mind: Reflections on Language and Music*. Shimla: Indian Institute of Advanced Study.

Mukherji, N. 2001. Shifting domains. In *Basic Objects*, A. Raina, and M. Chadha (Eds.). Indian Institute of Advanced Study: Shimla.

Mukherji, N. 2003. Is C_{HL} linguistically specific? *Philosophical Psychology* 16 (2): 289–308.

Mukherji, N. 2010. *The Primacy of Grammar*. Cambridge: MIT Press.

Quine, W.V.O. 1953. On what there is. In *From a Logical Point of View*. Cambridge: Harvard University Press.

Penrose, R. 1994. *Shadows of the Mind*. Oxford: Oxford University Press.

Penrose, R. 2001. Conversation. In *Mind, Matter and Mystery: Questions in Science and Philosophy*, R. Nair (Ed.), 119–133. New Delhi: Scientia.

Chapter 5
The Sceptic and the Cognitivist

> *There are inevitably going to be limits on the closure*
> *achievable by turning our procedures of understanding on*
> *themselves.*
>
> Thomas Nagel

This chapter is situated in a specific, alarming and disturbingly popular intellectual context. I will illustrate this context with two quick examples. Sometime during the mid-1980s, I happened to listen to a lecture by Roger Schank on how the techniques of strong artificial intelligence were beginning to capture some of the fundamental features of what was thought to be human common-sense reasoning, such as the decision to carry an umbrella when it is cloudy outside.[1] The theoretical merits of Schank's proposal need not detain us. During the discussion, someone raised some issues from the work of Immanuel Kant. Schank responded by saying that philosophical discussion of these issues are often too vague to be of any theoretical (=scientific) interest.

Soon after this talk, Patricia Churchland published a book interestingly titled *Neurophilosophy: Toward A Unified Science of the Mind-Brain* (Churchland 1986). In this book, Churchland took a breathtakingly rapid overview of the histories of epistemology, philosophy of mind and philosophy of science to conclude that, insofar as specific understanding of the human mind in its operational details is concerned, philosophy is hardly any guide; hence, the suggested shift to neuroscience. Again, the merits of Churchland's arguments need not concern us (see Mukherji 1990). What is of immediate interest is that this book turned out to be an unprecedented publication event. In general, it was widely embraced by the scientific community and also by much of the philosophical community.

This is a revised version of a paper published as Mukherji (2003).

---

[1]Roger Schank, Special Lecture, Society for Philosophy and Psychology, Annual Meeting, University of Toronto, October 1985.

© Springer Nature Singapore Pte Ltd 2017
N. Mukherji, *Reflections on Human Inquiry*,
DOI 10.1007/978-981-10-5364-1_5

5.1   Arrogance of Science

These examples have been chosen with some care; needless to say, there are many others. The work of Schank and Churchland has been widely influential in setting the agenda for cognitive science for some time, especially during 1980s and 1990s. Moreover, within cognitive science, Schank and Churchland had almost diametrically opposing views regarding the explanatory goals of cognitive science. While Schank may be thought of as an influential representative of the classical artificial intelligence community, Churchland favours the more recent connectionist and neural net frameworks. It is interesting, therefore, that they hold a very similar attitude of impatience with respect to classical philosophy.

Since the publication of Churchland's book, there has been a flood of influential publications by eminent philosophers, physicists, biologists, neuroscientists and cognitive psychologists (Dennett 1991; Edelman 1992; Penrose 1994; Chalmers 1996; Pinker 1997; etc.) which basically argue for the agenda outlined by Schank and Churchland to the effect that the time has come to jettison issues in human cognition from classical philosophy and to link it with current science. I must add that although this is the general message embodied in the publications just cited, the claim, unlike in Schank and Churchland, has not always been made explicitly. In fact, one finds a rather charitable view of philosophy in, say, Daniel Dennett and David Chalmers, being philosophers themselves. However, one is left wondering about the role of philosophy—except as a pleasing pastime for the clever minds— once their elaborate scientific programme unfolds.

In a way this attitude has an immediate appeal for the frustrated students of philosophy. After all, we do not find—say in Kant, Hume or Descartes, not to speak of Plato and Aristotle—any detailed explanation of a single cognitive phenomenon, though general intuitive clues abound. In some cases, say in some of the writings of George Berkeley (1709), we do find attempts to explain particular phenomena in some detail, but it is not at all clear how such empirical attempts mesh with Berkeley's general philosophical views. In other words, when Berkeley studied the Moon phenomenon or Dr Barrow's paradox, he was essentially adopting a scientific posture that was possibly delinked from his more pronounced philosophical ones. In any case, even these rare and piecemeal attempts at explaining actual cognitive phenomena have never been expressed in a theoretical fashion in which clearly defined variables, each with empirical significance and ranging over delineable classes of cognitive phenomena, are put together in a formal, deductive framework. So, one cannot glean even a rudimentary science of cognition from these attempts.

Things didn't change much even with the advent of modern logic because by then philosophy had come to be identified with conceptual analysis, rational reconstruction, linguistic clarification, and the like. Philosophy was not seen as explaining any aspect of our lives at all; that task was left for some future psychology. Thus, the first—and possibly the last—attempt at a philosophical examination of perception with the tools of modern logic, by Rudolph Carnap (1932), cannot be seen as offering a theory of perception, although much formal and conceptual ingenuity went into the construction of this edifice, for whatever it was worth. As it so happened, even this attempt at a conceptual-formal reconstruction of colour perception suffered from fatal conceptual flaws (Goodman 1951; Quine 1953). In general, it was obvious that although the concerns of Frege, Russell, Carnap, Wittgenstein and others did clarify various things dear to the philosophical tradition, things didn't quite progress beyond the generalities already available from Aristotle, Kant or Descartes insofar as understanding the character of human cognition was concerned.

Philosophers are traditionally comfortable with generality; it is the specificity of Ramu or Chotu's behaviour that makes them uneasy. So if we are wondering why little Babu keeps saying 10, 11, 13, 14 while regularly missing 12, or why Ramu cannot distinguish between the lower case *b* and the lower case *d*, or why they routinely miss themselves in counting the number of people in a room—all the philosophies of language, mind and mathematics aren't likely to be of much help. So, it is unsurprising that people like Schank and Churchland aren't impressed with the vague talk of schematism, categories, exemplars, and 'clear and distinct' ideas when they are interested, for example, in the exact details of how children solve the Tower problem, or in the exact sequence of events inside the head which ultimately results in its rotation towards a banana.

5.2 State of Philosophy

Let me try to understand quickly just what is at stake here. Philosophy, as we know it now, can no longer say useful things about space, time, motion, living forms, and the like. All of that, and much else besides, already belong to the various natural sciences. Similarly, various classes of abstract objects such as numbers, sets, figures and fields belong to various mathematical fields which now constitute the broad discipline of formal logic, given what is published in the current issues of *Journal of Symbolic Logic*. Regarding mental phenomena, qualitative mental phenomena, such as feeling of pain, emotional and other 'psychic' disorders, perception of various sorts, etc., all belong to assorted branches of psychology or of neuroscience or some combination of the two. By now, the erstwhile neighbourhood discipline of cognitive psychology has been virtually subsumed under the discipline of cognitive neuroscience, very far away from the departments of philosophy.

To recall, much of the human cognitive phenomena—in particular, phenomena usually dubbed as 'higher cognitive function' such as long-term memory, language, thinking, problem-solving and the like—were primarily philosophical concerns until very recently. Notwithstanding what we call these concerns—epistemology, philosophy of mind, philosophy of language, or whatever—the discussion of these domains had an unmistakable philosophical flavour. Even the avowed anti-metaphysical and proto-science proclamations of the logical positivists could not change things significantly.

So, in a way, the thinking mind was the last frontier that unquestionably belonged to philosophy. Now, if, via Schank and Churchland, that frontier has to be given up, the only domains that seemingly belong exclusively to philosophy are our moral, aesthetic and spiritual lives, and the subdomains thereof, if at all. 'If at all' because even the study of human and nonhuman moral and aesthetic processes are beginning to fall under cognitive psychology, as recent publications in influential journals *Cognition* and *Behavioural and Brain Sciences* suggest (Juslin and Västfjäll 2008; Cushman et. al 2013). Therefore, for those of us in the minority who do not believe that we can ever have interesting theoretical understanding of our spiritual life, philosophy, in the very near future may turn out to be a discipline without any substantive theoretical agenda. Incidentally, papers on religious practices have also started appearing in the cognitive science literature (Banerjee and Bloom 2013; Fincher and Thornhill 2012).

Obviously, I have taken a rather narrow view of what counts as doing philosophy. I have taken philosophy to be another discipline—a 'body of doctrine'—which gives an explanatory account of some realm of experience giving birth, thereby to science at opportune moments of its history; the expression *body of doctrine* is from the eighteenth-century chemist Joseph Black, cited in Chomsky (1999, p. 166) who in turn cites from Schofield (1970). To that extent, I have taken the old-fashioned 'mother of all disciplines' view of philosophy. The view is old-fashioned, but not obsolete. Jerry Fodor, for example, thinks that, for philosophers like Dewey and Austin, 'philosophy is what you do a problem until it becomes clear enough to solve it by doing science' (Fodor 1981, p. 177). What I have suggested above is that philosophy may be running out of such problems, if Schank and Churchland are right.

In a sense, the preceding forlorn observations might actually please one familiar form of thinking, especially in proponents of classical Indian philosophy, who deny that it is the task of (genuine) philosophy to generate understanding of human nature and the world. Philosophy, they claim, is a 'way of life'; so, the intricate *Nyaya* accounts of formation of veridical perceptual judgements, beginning with the contacts of the senses, are to be seen as ritualistic devices for attaining *moksha*. This is not the first time that avowed radicals seem to be joining hands with revivalists. We even have a nice word for that: *fundamentalism*.

There are, of course, other views of philosophy and other things that philosophers commonly find themselves engaged in. Thus, philosophers often get into the business of conceptual clarification of other bodies of inquiry; or, they clarify and sharpen various opinions held in common sense. It could be argued that these views of philosophy are largely immune to whether some classical philosophical issues get absorbed in science. If nothing else, philosophy prevails as usual by turning its attention to this new science in one of the role of philosophy just sketched, as the chapters in this volume arguably exemplify. Thus, philosophy continues to prevail as a second-order discipline after the emergence of cognitive science, just as it prevailed after the emergence of the earlier science, say, physics. Let us take a brief look, therefore, at the notion of philosophy as conceptual clarification, rational reconstruction, and sharpening of common sense with respect to the body of doctrine called *physics*. Does some sense of 'physical' philosophy prevail after physics branched off from philosophy centuries ago?

To begin with the last of the options, I do not think that we *hold* many common-sense views of the physical world outside physics. Either our prescientific views have already been sharpened by centuries of physics itself, or the few remnants of commonsense are quickly absorbed within the growing body of physics. The rest of the commonsense—for example, that things move only when pushed or that the Sun goes around the Earth—simply have to be given up even if they leave some philosophers unhappy. It is well documented by now that the 'mechanistic' philosophers in the Cartesian tradition never came to terms with the notion of action at a distance; Wittgenstein was said to have asked, what would it look like if it looked like as if the Sun is going round the Earth?

In sum, we do not seem to have an independent hold on the physical universe. Sometimes, of course, we hold views which are not immediately certified by the current physics. For example, it is reported that our prephysics view of the universe is largely Aristotelian in character as psychological experiments with children seem to suggest (Carey 1985). But that is not a problem *for* physics; that is a fact about our *thinking* which is precisely the domain under contention here. So much then for the activity of sharpening common sense.

As far as the rational reconstruction of body of physics is concerned, philosophy becomes essentially a secondary activity parasitic upon the development of physics. It is an exercise in historical pattern-recognition, which is ultimately based on some common notions of rationality, progress, unity, and the like. Again either we do not have enough theoretical control over these notions, or they belong to the study of mind and we are back to square one. If, on the other hand, rational reconstruction amounts to getting a summary view of physics where stable and important results are recorded in some facilitating order, then we are writing a *textbook* in physics; we are not doing philosophy.

Finally, for physics, it does not make sense to ask, following the method of conceptual clarification from outside of physics as it were, whether physics is on the right track, whether the body of physics is consistent, whether the concepts employed have empirical significance, whether the formalisms employed are backed up by a general theory, and so on. Such questions, of course, are often asked internally within the discipline of physics, and we may have a gallery view of observing, in such foundational exercises, some remnants of what we fondly thought were 'philosophical' concerns. But the fact remains that even if we may, on occasion, catch physicists doing some philosophy (say, in the continuing discussion on the measurement problem in quantum theory or the significance of the notion of dark matter); we are not doing it. I conclude that, insofar as physics is concerned there is nothing that we can do, except perhaps watch.

In my view a similar situation already obtains for the recent discipline of theoretical linguistics. As noted in Chap. 1 of this book and in a number of other essays, the study of the structure and function of language had been one of the central concerns of contemporary philosophy. Many aspects of these concerns have now been taken over by theoretical linguistics. The linguistic enterprise has already faced a number of what may be viewed as foundational problems. The conflict between descriptive and explanatory adequacies is one of the earlier ones. In the 1980s, the postulation of 'inner' levels of representation, d- and s-structures, posed another fundamental problem since no other system of the mind accesses them. In the current minimalist programme, such problems include the existence of uninterpretable features: lexical features, such as structural Case, that have no semantic interpretation are found in every language (Mukherji 2010). Notice that foundational problems have progressively become more theory-internal, as expected in an advancing science. It is not surprising that attempts to challenge the foundations of the discipline from the outside have more or less faded out in recent decades.

I spent a bit of time on physics and linguistics because these are the clearest example of what happens to philosophy when an offspring spreads its wings and leaves the nest. The situation isn't worrying as long as there are other offspring to bring forth and rear. As I read them, Schank and Churchland are telling us that the last of the philosophical progenies is raring to go. Just as everything we ever wanted to say about matter and motion has been absorbed in physics, everything we care to talk about language, thought and reality—except for the loose, dispensable parts of common sense—is going to be absorbed in cognitive science. From this futuristic hindsight, commonsensical explanations of our ordinary conceptions of knowledge and mental phenomena aren't likely to be of much value. Thus, '(t)he moral is that we ought to stop asking for analysis: cognitive psychology is all the philosophy of mind we are likely to get' (Fodor 1981, p. 202).

5.3   The Sceptic Rises

The picture of a barren future for philosophy so far developed assumes that Schank and Churchland are largely right in claiming that cognitive science is about to take over issues concerning mind and knowledge. We saw that if these issues are to lead us to grand theoretical moves which explain sundry behaviour of everyday life, then traditional philosophical ways of addressing these issues better be abandoned.

But are theoretical moves of such generality and specificity going to be available in the near future? Are they going to be available at all? Let us call these questions—especially, the last one—sceptical questions. In a way, from the point of view of cognitive science, such questions are intrinsically uninteresting. How can we tell now what an enterprise is going to look like in the future? Such questions could have been, indeed must have been, raised at the beginning of physics. But physics progressed, through calm and stormy times, without ever directly answering these questions. The questions were ultimately answered indirectly by the growth of physics itself to the point that scepticism in this domain became uninteresting. Therefore, Schank and Churchland will simply ask for more time.

Sceptical questions thus are not theoretical questions. In that sense, they are political questions. The sceptic is a political person in the sense that he raises his questions in advance of any determined theoretical activity since he is interested in questioning the very salience of that activity. He is not intervening in the (emerging) theory or suggesting alternative theoretical options. He is claiming, before the onset of science, that we possibly cannot overcome our state of ignorance. In that sense, the sceptic is a romantic and quite often a mystic. He is dazzled by the richness of human experience, he is overwhelmed with the unbounded complexity of the cosmos, and he is happy to stay that way (Mukherji 2006; *this volume*, Chap. 11). Scepticism thus is not a position, but a state of mind which is reluctant to submit to any position at all in the face of the totality of human experience.

This state of mind, of course, is best maintained in silence. As many authors on scepticism have pointed out (Strawson 1985), articulation of scepticism—in contrast to, say, a mere shaking of the head—is very nearly self-defeating. If the sceptic is pointing at some loopholes in a certain theory, he is merely suggesting counter-examples or raising logical or conceptual difficulties. In each case, he is asking merely for theory-change which is an activity, as suggested above, consistent with internal scientific practice. If the sceptic is raising more fundamental questions on the very nature of a scientific enterprise, then he must offer an alternative conception of the domain under review. In effect, he ends up offering his own theory, however loose, and freezes on an alternative standpoint. So sceptical arguments, when appealing, are more like gestures at erasure, rather than attempts at construction. We must then view the sceptic with empathy; he makes his presence felt somehow with inadequate verbalizations of an essentially silent protest. The sceptic is the subaltern of the otherwise elite academia. That is always the hallmark of real politics, such as class war (Spivak 1988).

It may be interesting, from a sweeping historical point of view, to look at each instance of the growth of theoretical knowledge essentially as the result of such politics. In particular, I suggest that the growth of theoretical knowledge be viewed as a political struggle between ongoing philosophy and some emerging science with the philosophical sceptic playing an essential role in that struggle. Philosophy, at time T_1, displays a certain general body of knowledge in a certain domain D which the emerging science wishes to take over at T_2. Between the interval T_1 and T_2, the sceptic attempts to neutralize the emerging science with a degree of success and a degree of failure. The end product is a new body of knowledge at T_2 which extends a bit beyond philosophy at T_1, but doesn't quite get to where science wished to go.

If there is enough continuity in scope, method, general style of argumentation, citation sources, etc., between philosophy at T_1 and the new product at T_2, we are still doing philosophy; if not, we have a new science. We aim at revolutions, but we settle for an evolutionary step of varying size. The labels *philosophy* and *science*, therefore, are not of durable interest; what is of interest is the role of the sceptic in restricting the scope of the new science at a level lower than its original expectations. Two caveats are needed at this point to lend some empirical flavour to the picture just sketched.

First, there may be factors internal to science (with its social setting) itself which restrict its development beyond a point during a certain period. The science might await developments in technology or in theoretical tools; there could be more direct social opposition, good or bad, to particular issues in science. Examples for each of these cases abound in the history of science. So, a temporary halt in the progress of science need not be thought of as a victory for the sceptical-philosopher. The specific form of struggle I am anxious to examine here concerns the encroachment of science in some domain of human experience where certain philosophical understanding has already been reached. The question then is whether the domain in its entirety has been absorbed in science or not.

A related caveat is that the picture sketched above does not rule out the possibility that in some domains, the philosopher-sceptic might always be at a losing end, historically speaking. The advent of modern physics, as we saw, is a clear case in point; the development of physics ultimately left no room for any residual philosophical understanding in this domain essentially because scepticism in this domain could only be of the self-defeating variety. It is not even clear what the form of 'silent protest' would mean in this domain. This is essentially the point Willard Quine (1969, p. 303) made from a somewhat different direction:

> (T)heory in physics is an ultimate parameter. There is no legitimate first philosophy, higher or firmer than physics, to which to appeal over physicist's heads. Even our appreciation of the partial arbitrariness or under-determination of our overall theory of nature is not a higher-level intuition; it is integral to our under-determined theory of nature itself, and of ourselves as natural objects.

Just to keep the record straight, I now think that I underplayed this powerful idea in my youthful Mukherji (1983). Nevertheless, I still think that Quine's idea applies exclusively to physics, and not to the entire discourse of 'natural objects'.

A vastly different picture emerges when we 'turn inwards', so to speak, and begin to address the domains of human mind and language themselves. There is an old adage that a theory of language is an impossibility since the theory has to be stated in some language or other. Thus, the theory always falls short of its object. It quickly generalizes to a dim view of theories of mind as well: a theory of mind is an impossibility since the theory itself will be a product of the mind, and hence a part of the object under examination. The adage appeals to the image of a spectacle: we can give only a partial and distorted description of the spectacle when we wear it; we can take it off, but then we cannot see. This adage is distinct from classical scepticism that denies the possibility of any knowledge. The effect of the adage is restricted only to cognitive inquiry; in that sense, it allows the possibility of knowledge of the 'external world,' say, the world of physics.

The adage is too general and vague to allow a direct critical examination. Hence it is seldom directly addressed; similarly, it is not directly addressed here either. The adage fosters a lingering intuition that our ability to have a theoretical grasp of ourselves must be severely restricted somewhere: 'There are inevitably going to be limits on the closure achievable by turning our procedures of understanding on themselves' (Nagel 1997, p. 76). McGinn (1989, p. 350) explicitly formulates the principle of cognitive closure:

> A type of mind M is cognitively closed with respect to a property P (or theory T) if and only if the concept-forming procedures at T's disposal cannot extend to a grasp of P (or an understanding of T).

It is likely that when we approach that point our theoretical tools begin to lose their edge and the enterprise simply drifts into banalities since, according to the adage, our resources of inquiry and the objects of inquiry begin to get hopelessly mixed up from that point on.

Such a point could be reached in the 'hard sciences' as well when they attempt to turn 'inwards'. This may be one way of understanding the origin of the deep puzzles around the so-called 'measurement problem' in quantum physics (see *this volume*, Chap. 2). The conjecture here is that, for inner domains such as reasoning and language, such points show up sooner rather than later. It is clear that no advance notice can be given for the occurrence of that event; the inquiry breaks down under the 'weight of its own contradiction', to use a Marxist locution. In terms of the discourse suggested here, the adage can only be entertained in silence with a look of incredulity at any enterprise that attempts to overcome its effect.

If we buy at least some of the elements of the sceptical perspective sketched above, we might get some hold on some puzzling features of the history of thought on mind and knowledge. Consider Kant, for instance. We are often told that Kant may be viewed as a philosopher who directly confronted the sceptical challenge to the very possibility of human knowledge with an ingenuous finesse. Instead of confronting the sceptical arguments directly—say, those posed by David Hume—he simply assumed that we have knowledge, and proceeded to describe the conditions under which empirical knowledge could be garnered. The very fact of knowledge then, with its underlying structural conditions, could be hoisted to silence the sceptic. As opposed to the sceptic, Kant is to be viewed as a 'constructionalist' (George 1982); he is going to awaken philosophy from its dogmatic slumber. In particular, he was going to tell us how physics and mathematics are possible. What did he construct in fact?

Generations of students of Kant are disappointed in not finding in Kant any thorough explanation of any interesting item of either physics or mathematics beyond scattered remarks on Newton, and banality of things such as '2 + 2 = 4'. Throughout his monumental work *Critique of Pure Reason* (Kant 1787), we find Kant wavering between two opposing pulls: (a) meeting the sceptic and justifying various sorts of knowledge we do have; and (b) giving a general account of such knowledge as part of his enterprise of justification. In other words, we see Kant as proposing some general account of, say, empirical knowledge basically to tell the reader that an account of that sort can be given. Kant is telling us that, contrary to the articulate sceptic, a theory of human knowledge is possibly at hand. As we saw, the articulate sceptic cannot consistently object to the program.

But the knowledge Kant ultimately gave an account of looks nothing like the theories of physics and mathematics he sought to construct. Thus, behind the scene, the sceptic did ensure his share of no-theory. In that way, I think that the profound end product of Kant's work—his categories, schemata, general principles, transcendental deductions, and the like—is laced with irony. The general picture of human knowledge outlined in the *Critique* is the end product of a political struggle at best half-won. As Noam Chomsky remarked in a slightly different context, the psychological generalizations reached by Kant's method of transcendental argument were 'evaluated by reflection, not empirical inquiry, and opens no research programmes—except, perhaps, into ordinary use of terms and concepts of rationality' (Chomsky 1999, p. 166). Do non-transcendental methods of empirical inquiry on language and mind fare any better?

5.4 Politics of Language-Theory

The very brief discussion of Kant was designed to highlight two things: (1) it supplied an example for the nature of the political struggle I have in mind; (2) it also supplied some idea regarding how the struggle is likely to work in the specific domain of human cognition. Perhaps the sceptic has already lost the politics of matter and motion (but who knows?). What, nevertheless, is the current status of the politics of cognition? At this point it will be worthwhile to survey the entire field of cognitive science as it has developed in the last few decades. For obvious reasons of space, I propose to restrict the survey to some brief comments on the developments in language-theory. However, I do not think that we are losing the generality of the discussion by this choice. Let me explain.

By some influential accounts (Chomsky 1991), current cognitive science, including language-theory, emerged in the mid-1950s out of developments in computability theory, gestalt psychology, logical theory, and certain branches of epistemology. Since then language-theory, especially the discipline inaugurated by Noam Chomsky, continues to be perhaps the only major success in this field. Another tested success theory is the theory of vision due to David Marr (1982) and others. However, this theory doesn't quite belong to the domains of mind and knowledge since, according to Fodor (1983), the visual system is an 'input system' while language, thought etc. are 'central systems.'

Returning to language-theory and cutting a very long story short, it is by now reasonable to conclude that near-fatal problems confront any attempt to extend the scope of language-theory beyond its initial domain of grammar (Fodor and LePore 1994; Mukherji 2010, Chaps. 3 and 4). At some point in the development of language-theory—while continuing with the pretheoretical notion of language which falls in the domain of philosophy—we must face the question of how linguistic/grammatical knowledge relates to world-knowledge. It is obvious that some theory of human conceptual system is needed to set up the link (Fodor 1998). By most accounts, a theory of meaning thus is nothing but a theory of concepts (Horwich 1998). So, roughly, problems regarding attempts to extend grammatical theory to include a study of meanings and problems regarding attempts to give a satisfactory account of non-linguistic world-knowledge will have large overlaps (Jackendoff 1983, 1990). Thus, if we have some idea of how the politics of cognition works in the case of language-theory, we might get an insight into the politics for the rest of the field.

In the early 1950s, Noam Chomsky thought that there must be a way of describing languages which focuses on certain universal features of language acquisition and use. We know, among other things: (a) humans are the only organisms that display the ability to acquire and use languages in the form of unbounded sequences of interpretable symbols; (b) language acquisition takes place under conditions of severe impoverishment of stimuli; and (c) children are not born with a specific choice of language since, given the right environment, they can pick up any of the innumerable human languages. It follows from these facts that humans must be endowed with some universal linguistic principles as part of their genetic make-up (Chomsky 1965, Chap. 1).

Notice that so far there are no restrictions on the concept of language except that one is concerned only with human languages, and not with metaphorical notions of language which may apply to dolphins, bees, birds, apes and DNA sequences (Chomsky 2000). To that extent, the notion of language under investigation is a commonsensical one, and is in line with thousands of years of philosophical reflections on this uniquely human phenomenon. Chomsky is clearly on record in tracing the lineage of his theory to classical linguistics such as Panini and Humboldt, and to philosophers such as Descartes, Kant, Wittgenstein, Austin and others.

However, as reflected in the very titles of Chomsky's two early books (Chomsky 1957, 1965), the initial theory was solely concerned with the syntactic character of human languages, and not with things such as word-meanings, semantic structures, usage, pragmatic conditions, speech-acts and the like. After several drastic modifications over the decades, the theory of syntax thus proposed has come to be regarded as the base theory for any theory of language; technically, a theory of I-language. For the purposes of this discussion, I will take the validity of this theory for granted.

It is important to note that the phenomenon of language had fascinated thinkers across traditions for well over a millennium, but thinkers before Chomsky never came close to explaining the phenomenon of language acquisition and language use even for the restricted domain of syntax, although we have had some rare examples of detailed description of some languages, notably that of Panini, in the past. Before Chomsky, all we had were either detailed enumeration of facts as in Panini or general remarks, often of a mystical nature, such as 'language is a totality of speech depositions', 'language is the house of being', etc. In that sense, we may view Chomsky's work as the first attempt towards a genuine scientific theory of language restricted, in the initial stages, to the study of syntax.

A huge debate immediately followed regarding the viability of Chomsky's project. It is interesting to take a quick overview of what happened. Linguists other than Chomsky, whose 'science' of language largely consisted in the study of speech-segments, found Chomsky's project to be essentially unscientific. Most philosophers and humanists, on the other hand, found Chomsky's study of languages to be too mechanical and devoid of the sensibility and the significance that animate our use of language; metaphors like 'bloodless' and 'lifeless' were routinely attached to Chomsky's conception of language (Mukherji 2010, Chap. 1).

Challenges of those sorts—though quite wide off the mark in some cases as Chomsky's torrential responses brought out over the years—only helped crystallize the study of syntax.

Skipping much detail, Chomsky's syntactic programme, by the mid-1970s, was a well-entrenched scientific programme whose empirical significance covered a wide variety of data across a large spectrum of languages. By the late 1980s, after another dramatic shift in the conceptual framework for linguistic theory, Chomsky's approach to syntax covered, according to Newmeyer (1991), over 90% of the published material in this area. Interestingly, much of what is going on in the adjacent areas of phonology, psycholinguistics, neurolinguistics, computational linguistics etc. is beginning to be directly inspired by research in syntax. Very little is now heard of the sceptical philosopher who challenged Chomsky's general approach to language and, thus, challenged his syntax work.

But the sceptic, as usual, wins his share of no-theory behind the scenes. By about the mid-1960s, some people were enthusiastic enough about Chomsky's syntactic programme to move on to issues in semantics to the point that they started looking for semantic universals. Notwithstanding Chomsky's own position on these issues, people began formulating various versions of universal and generative semantics. The results, after about a decade of furious research, were close to disaster. Either the versions won't fit syntactic description or they would make completely unargued assumptions about the functioning of the mind or, usually, both. In trying to keep to syntax while accommodating some of the interesting data that the abandoned research on generative semantics unearthed, Chomsky came up with a theory —*Extended Standard Theory*—which had the usual syntactic components, but it also had an additional component called LF (for 'Logical Form') where much of the syntactically-sensitive semantic information begins to get clustered (Hornstein and Weinberg 1991).

Yet the fact remains that LF has nothing to do with semantics in the sense in which semantics spells out the relation between language and the world (Soames 1989). As students of logic can quickly figure out, LF has to do with things like quantifier movement, term dependencies, scope resolutions, and the like. An LF-representation of a string tells whether a string is structurally ambiguous, which lexical element depends on another element for interpretation, and so on. For example, differences in LF-representations exhibit why the string *birds that fly instinctively swim* is semantically ambiguous; *instinctively* can modify either *fly* or *swim*: birds either fly instinctively or swim instinctively (Berwick and Chomsky 2016, p. 8). Nonetheless, in an important sense, an LF-representation *does not* tell what the specific referential and conceptual meaning of the string *is*. For example, the computational system will generate identical LF-representations for the strings *John tried to attend college* and *John decided to attend church* (Mukherji 2010, Chap. 4).

In that sense, the *linguistic* notion of LF captures the idea of minimal semantics to adhere to the traditional notion of language as a device for sound–meaning correlation. In that sense, a grammatical theory is an 'incomplete' theory of language; it awaits processing by other components of the mind (possibly geared to language as well) to form the complete semantic representation of a string. Thus, if our notion of meaning is supposed to capture such things as what we think, what we refer to and what is (ultimately) communicated to a hearer, then grammatical theory needs to be supplemented with further theories in tandem with the naturalistic syntactic theory already reached.

The beauty of Chomsky's recent work lies in accommodating, within the well-tested syntactic format, some classical issues of logical syntax without getting into model-theory at all. In that sense, LF is nothing more than extended syntax. But it does something more than 'pure' syntax as it was conceived in the early phase of the enterprise. It does explain, within a unitary format, some facts about meanings without referring to meanings as such, as noted. To get a quick feel of how it works, consider a simple multiple-quantifier sentence such as *Someone attended every seminar* (Hornstein 1995, p. 155). It is obvious that the sentence is two-ways ambiguous, and we will want a language-theory to generate two interpretations of the sentence. An array of syntactic principles can now be invoked to generate two structures with different scopal properties from the given lexical items themselves (Mukherji 2010, 2.4). So LF does go a bit beyond syntax without going beyond what I am calling 'extended syntax'.

Let us call the object so identified by current linguistic theory *grammar*; hence, the title *grammatical theory*. I am thinking of grammar as a domain which is larger than the domain of pure syntax, but which falls far short of our intuitive domain of language. So, after wrestling with the sceptical challenges regarding the very possibility of language-theory for several decades, linguistic theory did indeed come up with a new object.

Sceptical challenges to the theory, of course, are still raised. In an influential work, Saul Kripke argued (Kripke 1980, 1982) following some ideas of Wittgenstein, that the very notion of linguistic competence that underlies Chomsky's work lacks foundations. It turns out, however, that the sort of examples Kripke uses to promote this claim—the differences between *plus* and *quus*—cannot be found in the domain of grammar, though they are eminently valid in the domain of rules which seek to relate symbols with what they stand for (Mukherji 2010, Chap. 4). In other words, the notion of semantic competence, if understood in a full-blooded sense, is probably beyond rational enquiry. But, as we just saw, grammatical competence—not just syntactic competence—is a perfectly valid object for theoretical enquiry. This point connects nicely with Stephen Shiffer's sceptical conclusion (Schiffer 1987) that if we are looking for a theory of semantics of propositional attitudes, then the best theory to settle for is likely to be a No-Theory theory (see *this volume*, Chap. 9).

Generalizing on these ideas, we will expect cognitive science to grow in areas where it is possible to identify grammar-like objects: a grammar of vision, music, face recognition, map-making, logical thinking, kinship, etc. It is also possible that hidden among these disparate grammars there is a unitary notion of grammar that picks out some of these domains. For attempts to locate this concept and its philosophical and empirical implications see Mukherji (2000, 2010). Yet, none of these are likely to be full-blooded theories of visual perception or of knowledge of music, and the like.

To conclude, then, let us return to Fodor's remark about cognitive psychology being the only philosophy of mind. My hunch is that abstract theories of 'grammar' are all the cognitive psychology we are ever going to get in any domain. We may generalize on Chomsky's remark that 'grammars have to have a real existence.' However, theories of 'language', embodying the real, pulsating human condition, may remain beyond human reach. The study of 'language' thus falls under no-theory. Behind the scenes, the philosopher-sceptic will always be needed to push wedges of ignorance between the cup of human experience and the lips of arrogant science.

References

Banerjee, K., and P. Bloom. 2013. Would Tarzan believe in God? Conditions for the emergence of religious belief. *Trends in Cognitive Science*, 17(1): 7–8.

Berkeley, G. 1709. An essay towards a new theory of vision. In *The Works of George Berkeley: Bishop of Cloyne*, 9 volumes, A.A. Luce and T.E. Jessop (Eds.), 1948–1957. London.

Berwick, R., and N. Chomsky. 2016. *Why Only Us: Language and Evolution*. Cambridge: MIT Press.

Carey, S. 1985. *Conceptual Change in Childhood*. Cambridge: MIT Press.

Carnap, R. 1932. *The Logical Structure of the World*. Translated by Rolf George. Berkeley: University of California Press.

Chalmers, D. 1996. *The Conscious Mind. In Search of a Fundamental Theory*. Oxford: Oxford University Press.

Chomsky, N. 1957. *Syntactic Structures*. The Hague: Mouton.

Chomsky, N. 1965. *Aspects of the Theory of Syntax*. Cambridge: MIT Press.

Chomsky, N. 1991. Linguistics and cognitive science: Problems and mysteries. In *The Chomskyan Turn*, A. Kasher (Ed.). Oxford: Basic Blackwell.

Chomsky, N. 1999. *New Horizons in the Study of Language and Mind*. Cambridge: Cambridge University Press.

Chomsky, N. 2000. In *The Architecture of Language*. The Delhi Lecture, N. Mukherji, B.N. Patnaik, and R. Agnihotri (Eds.), New Delhi: Oxford University Press.

Churchland, P. 1986. *Neurophilosophy: Toward a Unified Science of the Mind-Brain*. Cambridge: MIT Press.

Cushman, F., R. Sheketoff, S. Wharton, and S. Carey. 2013. The development of intent-based moral judgement. *Cognition*, 127(1): 6–21.

Dennett, D.C. 1991. *Consciousness Explained*. London: Little Brown & Company.

Edelman, G. 1992. *Bright Air, Brilliant Fire*. New York: Basic Books.

Fincher, C.L., and R. Thornhill. 2012. Parasite-stress promotes in-group assortative sociality: the cases of strong family ties and heightened religiosity. *Behavioural and Brain Science*, 35(2): 61–79.

Fodor, J. 1981. *Representations*. Cambridge: MIT Press.

Fodor, J. 1983. *The Modularity of Mind*. Cambridge: MIT Press.

Fodor, J. 1998. *Concepts: Where Cognitive Science Went Wrong*. Oxford: Clarendon Press.

Fodor, J., and E. LePore. 1994. Why meaning (probably) isn't conceptual role. In *Mental Representation*, S. Stich, and T. Warfield (Eds.), 142–156. London: Basil Blackwell.

George, R. 1982. Kant's sensationalism. *Synthese* 47: 229–255.

Goodman, N. 1951. *The Structure of Appearance*. Cambridge: Harvard University Press.

Hornstein, N., and Weinberg. 1991. The necessity of LF. *Linguistic Review 7*, 129–167.

Hornstein, N. 1995. *Logical Form: From GB to Minimalism*. London: Basil Blackwell.

Horwich, P. 1998. *Meaning*. Oxford: Oxford University Press.

Jackendoff, R. 1983. *Semantics and Cognition*. Cambridge: MIT Press.

Jackendoff, R. 1990. *Semantic Structures*. Cambridge: MIT Press.

Juslin, P.N., and D. Västfjäll. 2008. Emotional responses to music: The need to consider underlying mechanisms. *Behavioral and Brain Sciences* 31(5): 559–575.

Kant, I. 1787/1929. *Critique of Pure Reason*. Translated by. N.K. Smith. London: McMillan.

Kripke, S. 1980. *Naming and Necessity*. Cambridge: Harvard University Press.

Kripke, S. 1982. *Wittgenstein on Rules and Private Language*. Cambridge: Harvard University Press.

Marr, D. 1982. *Vision*. New York: Freeman.

McGinn, C. 1989. Can we solve the mind–body problem? *Mind* ICVIII, 391 July, 349–366.

Mukherji, N. 1983. Against indeterminacy. In *Humans, Meanings, Existences*, D. P. Chattopadhyay (Ed.). New Delhi: Macmillan & Co.

Mukherji, N. 1990. Churchland and the talking brain. *Journal of Indian Council of Philosophical Research* (August).

Mukherji, N. 2000. *The Cartesian Mind: Reflections on Language and Music*. Shimla Indian Institute of Advanced Study.

Mukherji, N. 2003. Skeptical politics. In *Truth and Value: Essays in Honour of Pabitra Kumar Roy*, R. Ghosh (Ed.). New Delhi: New Bharatiya Book Corporation.

Mukherji, N. 2006. Textuality and common life. In *Literature and Philosophy*, S. Chaudhury(Ed.). Papyrus: Kolkata.

Mukherji, N. 2010. *The Primacy of Grammar*. Cambridge: MIT Press.

Nagel, T. 1997. *The Last Word*. New York: Oxford University Press.

Newmeyer, F.J. 1991. Rules and principles in the historical development of generative syntax. In *The Chomskyan Turn*, A. Kasher (Ed.). Oxford: Basil Blackwell.

Penrose, R. 1994. *The Shadows of Mind: A Search for the Missing Science of Consciousness*. Oxford: Oxford University Press.

Pinker, S. 1997. *How The Mind Works*. New York: Penguin Books.

Quine, W.V.O. 1953. *From A Logical Point of View*. Cambridge: Harvard University Press.

Quine, W.V.O. 1969. Reply to Chomsky. In *Words and Objections*, D. Davidson, and J. Hintikka (Ed.). Dordrecht: D. Reidel.

Schiffer, S. 1987. *Remnants of Meaning*. Cambridge: MIT Press.

Schofield, R. 1970. *Mechanism and Materialism*. Princeton: Princeton University Press.

Soames, S. 1989. Semantics and semantics competence. *Philosophical Perspectives* 3: 575–596.

Spivak, G.C. 1988. Can the subaltern speak? In *Marxism and the Interpretation of Culture*, C. Nelson, and L. Grossberg (Eds.), 271–313. Macmillan Education: Basingstoke.

Strawson, P. 1985. *Skepticism and Naturalism: Some Varieties*. New York: Columbia University Press.

Chapter 6
From Things to Needs

> *The philosopher labours to produce a systematic account of the*
> *general conceptual structure of which our daily practice shows*
> *us to have a tacit and unconscious mastery.*
>
> Peter Strawson

6.1 Introduction

One of the distinguishing features of totalitarian systems is that they seldom promote active philosophical traditions as part of their high culture. I am viewing totalitarian systems—roughly as it is described in Karl Popper (1962)—as closed systems of *thought* promulgated by some authority. In many ways, this view is different from, say, Noam Chomsky's view of totalitarianism as a system of tyrannical *organization*, although the two views coincide for prominent joints of history. Chomsky's list consists exactly of Bolshevism, Nazism and Corporations. My list includes these and extends to most of religious cultures, tribal cultures, etc., though many of these cultures need not be viewed as tyrannical. In fact, forms of tribal culture may open the avenue for human survival precisely by dint of their 'dogma'.

Sometimes totalitarian systems do advertise official philosophical doctrines, but these are typically instruments of propaganda, rather than vehicles of creative criticism. For example, the doctrinaire Marxist philosophy propagated in the erstwhile Soviet Union or in contemporary China, has very little to do with the radical, critical features of Marxist philosophy that originated in democratic Britain. In fact, apart from blocking off other philosophical traditions from the Soviet society, doctrinaire Marxism has done most damage to the living tradition of Marxist philosophy itself. Arguably, a similar picture attached to Buddhist philosophy as it became the official doctrine of the state at a certain stage of its development. We may be witnessing a similar phenomenon as versions of Buddhist philosophy,

---

This is a revised version of a paper published as Mukherji (2002).

© Springer Nature Singapore Pte Ltd 2017

N. Mukherji, *Reflections on Human Inquiry*,

DOI 10.1007/978-981-10-5364-1_6

nurtured around cult figures such as the Dalai Lama, increasingly become the favoured doctrine of a powerful section of the elites. However, the conservative, dogmatic character of much of Indian philosophy is a wholly different phenomenon to which we return.

In this sense, the presence of a lively and versatile philosophical tradition in a culture is a mark of its liberal character. Although every philosophical tradition attempts to reach a general, comprehensive understanding of the human condition, such understanding is typically routed through a critical engagement with other dominant systems of knowledge. For, given the richness and complexity of human experience, a comprehensive understanding can be approached only by relentless questioning of received positions. These positions include appeals to divine fore-knowledge, religious doctrines, science, or even common sense; a philosophical tradition questions each of these forms. A lively philosophical tradition is thus necessarily sceptical and heretic in character.

Several consequences follow even from this very brief sketch of the nature of philosophy (Mukherji 2003, *this volume*, Chap. 5 for more). First, the mere presence of philosophical thought is not enough to sustain a tradition. Second, a philosophical tradition is sustained only when it is able to engage constantly with other dominant forms of knowledge. As knowledge-systems become wider and more complex, the philosophical enterprise itself becomes progressively harder and sophisticated for such engagement to have any lasting value. Third, in order for a philosophical tradition to be significantly critical of others, it must develop tools and discourses to be able to critically examine its *own* edifice of knowledge. Constant self-examination, leading perhaps to self-rejection at times, has been a liberating feature of philosophy in any tradition since antiquity.

It is obvious that the conditions just stated could only materialize in institutional forms which are largely free from external control. Moreover, since a lively philosophical tradition, as noted, is *not* geared for promoting dominant systems of knowledge and practice, it fails to serve the interests of any major force in a society. Given this essential 'minority' status of philosophy, it follows that a philosophical tradition cannot be sustained unless the society as a whole is tolerant with enough space for the freedom of minority opinion.

It is well known that academic institutions of a certain liberal form have supplied those spaces in history: the Greek Lyceum, *Nalanda*, ancient seminaries, classical *gurukuls*, and, of course, the modern university system. The qualification 'liberal form' cannot be overstressed here. As mentioned, the very availability of an academic institution, or an institution of learning, is not sufficient for genuine philosophical activity to flourish. For example, the Soviet Union had an Academy of Philosophy; and Buddhist and *Vedantin* texts are routinely studied in monasteries and *mathas* respectively. But these are not exactly the places where one expects to find a lively philosophical dialogue, with the features stated above, to ensue; however, this is not to suggest that no useful work has ever emerged from these institutions. Philosophy thus depends on a rather thin margin of survivability; a liberal academic environment is likely to be its only habitat.

6.2 A Tradition 'Preserved'

Our attention is thus turned towards the character of philosophical practice in the academic institutions in India. As mentioned, there is no doubt that classical India did develop a variety of institutions which encouraged a lively philosophical tradition. The very fact that Indian philosophy branched off into a number of competing schools of thought, which questioned each other's foundational assumptions at great depth for well over a millennium, is an unmistakable sign of the presence of liberal academic institutions. The fascinating historical question of just which array of institutions and social forces made this achievement possible is seldom the subject of rigorous scholarship. Yet the sheer volume, range, quality and diversity of this work testify to the presence of a liberal mindset up to a certain point in time.

This is obviously a sweeping generalization which is in need of more careful and qualified formulation. For example, explanation is needed for the fact that the original sources of the *Carvaka* school of thought were first systematically obliterated, and then the school was subjected to one-sided denunciation. Yet the very fact that the *Carvakas* along with the Buddhists, Jainas and others were able to develop *at all* points to the abiding presence, over long periods, of what Amartya Sen (2001) calls 'intellectual heterodoxy'.

For a variety of ill-understood historical reasons, the classical system of institutions either fell apart or their continuing forms could not sustain the tradition as it was developed earlier. In an interesting little article, the Oxford philosopher Michael Dummett (1996, 14) traced much of this downfall to the 'massive impact of Western Culture… (which) has been all the more crushing because political hegemony accompanied cultural imperialism'. Textual evidence seems to suggest, however, that active, 'heterodox' philosophical activity essentially came to a halt many centuries before the British cultural invasion. It would seem rather that philosophical practice had *already* lost much of its vitality for it to resist, or to come to honourable terms with, Western 'cultural imperialism'. So, the real explanation here is likely to be more complex and less charitable than what Dummett proposes.

In any case, Dummett offered an interesting view, which is largely unaffected by questions regarding historical detail, on the *consequences* of this cultural invasion. 'As a result', Dummett observed, 'indigenous traditions have been, not killed, but *blanketed*' (emphasis added). By *blanketing*, Dummett means that 'the tradition did not die: it was, and still is, *preserved*' (emphasis added). The classical tradition 'was being handed down, without alteration, but not being added to; the creativity had gone (Dummett 1996, p. 15). When the instinct of preservation dominates a tradition, it begins to lose contact with the rest of the knowledge systems that subsequently arise by dint of the open-ended nature of human experience; this is what the term *blanketing* signifies.

Moreover, the instinct of (self-)preservation is directly opposed to any form of self-criticism, which, we saw, is one of the central features of a living tradition. In such a situation, Dummett pointed out, the tradition would no longer be interested in asking critical questions such as: 'Are the distinctions made correct distinctions?'

'Are there other distinctions which should have been made but have been blurred?' 'Are the arguments compelling?' And, ultimately, 'Are the conclusions true?' Dummett notes that only a philosopher, not a historian, would ask these questions. Therefore, when these questions are no longer asked, we have to acknowledge that the philosophical tradition has come to an end.

The net effect of these observations is that classical Indian philosophy never adjusted itself to what is now called *modernity* and the vast systems of knowledge it unleashed. As centuries passed and the scope of the 'blanketed' tradition became narrower, the tradition itself began to acquire features of obsolescence. It is natural, thus, that at least for the intellectual class, which was directly exposed to Western 'cultural imperialism', the knowledge-system enshrined in the tradition lost its intellectual appeal. In fact, in time, this class must have found this *sanskritised* tradition to be perhaps more alien than the classical traditions that formed the basis of *Western* knowledge-systems.

This last point raises difficult questions about the 'Indianness' of classical Indian philosophy and, by parity of reason, the 'Westernness' of Western philosophy. If a contemporary student of philosophy in India finds what is labelled *Western philosophy* to be more intellectually appealing than the frozen versions of Indian philosophy offered to him, does the student cease to be a part of the Indian tradition in any significant sense? I return to some of these questions in the final section.

6.3   Pundits and the Intellectual Elite

The considerations just raised are crucial for understanding the modern history of academic philosophy in India. In my opinion, many features of this history are routinely misconceived. For example, in a complaining tone, Dummett ascribed the lack of philosophical creativity to the fact that 'the intellectual elite did not participate in the process; they had studied philosophy at the universities, but philosophy written in Greek, or English, or German, or Latin, or French, but not in Sanskrit'. 'The philosophical formation', he contended, 'like the whole intellectual formation, was as it was because under the British Raj an alien educational system had been imposed, and, with it, an alien intellectual tradition and orientation' (Dummett 1996, p. 15).

So, the picture Dummett paints has all these pundits and scholars of traditional knowledge waiting in vain with yellowing texts in hand, but the 'intellectual elite' won't show up for lessons; instead, they ran to the universities to feast on Western philosophy. In this, the 'intellectual elite' is viewed as a servile and gullible lot who can be easily infected with an alien tradition which, like an overgrown tumour, ultimately destroys the parent body. If that were indeed the case, then the obvious prescription would be to enter into some surgical process to remove the alien structure such that, after a period of supervised nursing, the patient is able to return to the 'original' state.

Call it *Hindutva* or whatever, in effect it would mean that the philosophical practice in India should return to what the pundits preached, and that it should *stay* there. There is a growing voice, usually out of print, in the academic circles in India that this roughly ought to be the case. In Mukherji (1996), I have given some indication as to the sources and the character of the issue here. At least one influential voice, complaining about the 'exaggerated importance' to Western analytic tradition and suggesting major 'revision and updating of the syllabuses', is cited and examined there. Given the massive presence of Western philosophy in curricula and elsewhere, it is perhaps impracticable, according to this view, to banish Western philosophy altogether. Yet, for the sake of national purity and indigenous initiative, at least some steps in that eliminative direction are urgently needed.

Interestingly, if by *Western philosophy* we mean Western *analytic* philosophy, which in fact we *do* in what follows, then the opinion just sketched is likely to be shared by many academicians in the general area of humanities and social sciences. They would rather endorse that part of Western philosophy which is commonly labeled *continental thought*. I will not examine this angle since it does not as yet have a very large following within academic philosophy, even though it has virtually saturated the 'philosophy'-side of the humanities and social sciences. In any case, practitioners of this angle in India, especially those outside of professional philosophy, are not known for their understanding of classical Indian thought beyond lip service. Their attention is squarely focused on the likes of Barthes, Foucault, Lacan, Derrida and others. Timely mention of *Vedanta* or the *Upanishads* seems to be just a matter of political convenience (Mukherji 2006b). At issue also are dictums such as 'the enemy of an enemy is a friend'. Those who refuse to follow should be viewed as agents of Western culture.

For now, I am concerned only with the factual and the philosophical basis of this opinion. So, I am ignoring other matters such as where this opinion is coming from, how it is sought to be put into practice without any debate at all, the location of this opinion in the power hierarchy of academic philosophy, the lure of the growing market for a revivalist view of classical Indian thought in the West in the form of South-Asian Studies, the role of the Indian Council of Philosophical Research, and the like. Also, I am not suggesting that everyone upholding the revivalist view has a vested interest of the kind suggested above. Some obviously do; most are just confused about how to understand the complex relationships between issues of cultural identity and contents of professional disciplines in a post-colonial set-up.

Returning to Dummett, this popular charge of almost a moral failure of the intellectual elite needs to be assessed with care. For, if the charge was valid *then*, it ought to be largely valid even now since nothing much has changed; in fact, the scale and power of Dummett's 'intellectual elite' has increased manifold in the meantime with their deep penetration in the establishment. So, if what Dummett complained about is the case, it would be difficult to find a way out of *mass* moral failure.

Setting aside the wider issue of the 'whole intellectual formation' raised by Dummett, can we trace the widespread penetration of Western philosophy in the Indian academic scene wholly to the *imposition* of an alien educational system? As a matter of geographical fact, the educational system so introduced was no doubt alien. Also, we need not ignore the vile politico-cultural motivations, if any, for introducing this system. Let us grant as well, as a matter of fact, that the intellectual class, that jostled for the fruits of occidental culture in droves, basically grew out of this educational system; some of them might even have shared the underlying politico-cultural motivations, if any.

Yet these assumptions just do not explain the *unique phenomenon* of the entry and practice of Western philosophy in India at such a scale. The factors listed above must have existed at many places around the globe as the British Raj set about its Sun-following mission. Similar phenomena must have accompanied the *French* Raj in other parts of the globe, and, as everyone knows, French political hegemony is even more directly associated with Eurocentric 'cultural imperialism' . But the fact remains that in the last century Western philosophy found no lasting foothold anywhere else in the non-Western world *other than India* (unless we wish to think of Australia and New Zealand as non-Western).

More importantly, the proponents of the 'imposition' view need to explain the following widely attested facts. First, the resistance to the British-imposed educational system, cultural imperialism, and to the British Raj as a whole basically ensued from the Western-educated classes itself. The *traditional* Indian elites, largely dominated by a section of the Brahminical class, were generally not distinguished on that count. Second, in contrast to some of the classical acts of the orthodox Hindu society, there is no tangible record of direct imposition of the Western ethos in terms of, say, destruction of texts or of centres of learning. If anything, the evidence points to the opposite. Given the limited intellectual calibre of the actual colonizers, there was some effort in continuing with the *preservation* of traditional culture in terms of opening of libraries, archives and colleges dedicated to the pursuit of traditional knowledge: the array of Hindu and Sanskrit colleges across the country testify to that effort.

There are thus grave doubts as to whether the political hegemony in fact *wanted* the educational system to foster modernity in the true sense. To believe in *that* is to entertain the naïve belief that Western imperialism would in fact be interested in creating another Eurocentre out of the wilderness of Asia after the loot is over.

The reason why I am directing attention to Michael Dummett, rather than to the omnipresent *Hindutva* advocate in the Indian scene—and their closet colleagues in the allegedly liberal academia advocating *Swaraj* in ideas—is politically obvious. Dummett is one of the major post-war (analytic) philosophers in the world. Apart from significant contributions to many technical areas of analytic philosophy, his career as a teacher at Oxford University helped sustain a long tradition of liberal excellence practiced there. Apart from his philosophical presence, Dummett is also widely known for his work in support of the immigrants in particular, and against racial discrimination in general. There is no measure, therefore, with which he could be identified with the *interests* of *Hindutva*. His disinterested opinion thus supplies a powerful plank for the *Hindutva* ideologues to spring from. That is why it is important to show that Dummett's explanation of why Western philosophy took firm roots in India is, at best, simplistic; at worst, it is plain wrong. A more natural explanation of why Western philosophy entered the Indian academic scene on such a scale can be easily constructed if we are prepared to shift from perspectives such as Dummett's.

To begin with an obvious fact, even under the political hegemony and 'cultural imperialism' of the British Raj, Indian society, as a whole, never became a totalitarian system, although the space for active liberal practices was surely shrinking. Given the massive diversity of cultures upholding heterogeneity of thought and practice, there always were some liberal spaces for the intellectual class to occupy and explore. So, for sections of this class—which really is the only class at issue here given that academic philosophy has always been a part of high culture— engagement with a philosophical tradition was clearly a lively option. This ought to be especially true for a class whose ancestry goes back to a profound *indigenous* tradition within living memory. This last point alone distinguished the Indian scene from several others which came under the British or the French Raj.

Yet the domestic tradition which was currently available was a blanketed one. By then, centuries of acts of preservation and blanketing had led to a situation where critical thought had been replaced with a series of mindless rituals. There was a strong emphasis on restricted lifestyles, long and demanding religious practices, the accurate memorization of whole texts, great fuss over mastering 'pure' Sanskrit, absolute loyalty to the teacher and the tradition as *he* represented it, winning of open 'debates' with contrived hair-splitting arguments just to score points over the opponent and to impress the gathering, and the like. Needless to say, these practices were laced with a reverence for the caste system and with downright reactionary views about other cultures, women, and lesser mortals. As noted, these practices were at once the source and the consequence of features of obsolescence that infected large areas of philosophical *thought* itself. It is unlikely that liberal sections of the intellectual class would have found it appealing to engage with the listed modes of thought and practice.

In my opinion, it is rather important to raise and understand this scenario without any moral stick in hand: no individual or group is to be *blamed* for these happenings. On the one hand, the pundits and their disciples performed the salutary service of preserving the tradition for centuries against heavy odds. Scholarly documentation of their lives is hardly available. Yet, from what one can glean from some of their well-known twentieth-century representatives, no tribute seems adequate. Satyajit Ray, in his memorable cinema *Pather Panchali* (Song of the Road), portrayed such a character to highlight the general impoverishment of this aspect of the culture. However, to my knowledge, this aspect of Ray's film has never been discussed with sufficient historical material.

In sharp contrast to the self-serving image of the current university-based academician, these pundits typically led a difficult life with unflinching devotion to scholarship and erudition. As subsistence allowances from the state dried up, most of them were compelled to take up the profession of *purohit* (priest) to be able to maintain their families. This required long journeys on foot and indiscriminate fasting for a meagre and uncertain package of money, rice and *dhoti*. In order to survive, the self-demeaning character of this lifestyle, which some of our outstanding scholars had to endure, was wholly internalized in the pundit culture to the point where the tradition of *thought* itself was sought to be identified with this structure of impoverishment. As a result, orthodoxy and ritualism inevitably seeped into philosophical thinking.

On the other hand, the liberal sections of the intellectual class can hardly be blamed for shying away from the tradition represented by the pundits. Not without reason and from what they saw, much of the ills of the society around them, including the restrictions on open-ended rational inquiry, could be traced to the knowledge-systems represented by the pundits. During this tragic period in Indian history, the Western philosophical tradition, supported by the complex system of universities, developed in leaps and bounds in close contact with Western science. As the British- and missionary-sponsored education system supplied access to the English language and whiffs of Western high culture, the intellectual class in India rushed to Western philosophy as ducks to water.

Although it is seldom noted, it stands to reason that this quick and mass transfer to Western philosophy cannot be fully explained from the fact of 'liberal urge' alone. As noted, the classical *Indian* tradition itself nurtured a vigorous discourse of rational, analytical inquiry for over a millennium (Mohanty 1992; Sen 2005). Although the practicing part of the tradition had lost much of this analytical character, the memory of the discourse was still clearly documented in the texts and the commentaries. Moreover, much of the mode of rational inquiry must have survived in the *general* culture due to the continued, albeit diminishing, heterodox character of Indian society. It is just that the official practitioners, the pundits, failed to uphold this character in explicit terms. In sum, there must have been close, internal links between the *mindset* of the Indian intelligentsia and Western philosophy for the latter to attract the former.

Let me try to bring out this point from another direction. It is well known that Christian missionaries of various hues were present in India well before the advent of the British. Thus, a large system of churches had existed in India for centuries; the British Raj just enlarged the process manifolds. With these churches and related cultural and educational institutions, Western classical music must have had a substantial presence in India. It will not be surprising if the scale of 'imposition' of this music was even larger than that of the education system Dummett mentioned.

Yet, even today, this form of music never took roots with the liberal intelligentsia in India. Classical music in India continues to be decisively Indian in character. The reasons are not far to seek. For one, Indian classical music, with its artful symbiosis with Islam centuries ago, continues to be a *living* tradition in the sense under discussion here. So, there was no *need* for a mass shift in culture. For another, despite the alleged universality of music, the post-Bach musical tradition of the West with its harmonies, counterpoints and orchestral structures is markedly different in spirit, style and content from the Indian tradition to allow spontaneous formation of close links between the two. For philosophy, however, the fact of seamless integration clearly suggests that there was no such watershed of cultures between classical Indian philosophy and modern Western philosophy.

The perspective just proposed can be further substantiated with some actual case studies. Consider, for example, the work of Prof. K. C. Bhattacharya. As Arindam Chakravarty (1996, p. 1) remarks, this reclusive and outstanding scholar of *Vedanta* could well be regarded as 'by far the most original, subtlest and toughest of all twentieth-century professional philosophers in India'. As Chakravarty proceeds to elaborate:

The chapters on bodily subjectivity in his major work *The Subject as Freedom* anticipates some of the finest insights of Phenomenology. His *Studies in Vedantism* as well as the classic essay 'The Concept of Philosophy' allude to Kant's ideas on thinkability and knowability, albeit in a sharply critical manner, as if Kant and Samkara were equal parts of India's intellectual traditions.

So here was a typical *Vedantin* scholar who was naturally drawn, on purely philosophical grounds, to study Kant, Hegel and others to *re*activate his own philosophical tradition. As a result, we find perhaps the only distinguished work in professional philosophy produced by an Indian academic in the twentieth century. Although creative work of this calibre is difficult to cite, many authors can be mentioned who basically attempted to follow a similar course of enquiry.

6.4 Projections

There is thus sound basis to a perspective in which much of contemporary Western philosophy may be seen to have an underlying continuity with much of classical Indian philosophy. If facts of geography and other embodiments are not to clutter our vision, one may even suggest that classical Indian philosophy simply 'shifted' to the West once its space was lost in India. Mukherji (1997) contains some hints towards a similar way of relating the ancient work of Panini and contemporary generative grammar in the discipline of linguistics. The spirit of high ideas knows no boundary. If this view is even partially valid, then it follows that contemporary Western philosophy in India is a practice that continues the classical Indian tradition, albeit with a detour via the West. It is unclear if the same could be said of *the pundit-style practice of Indian philosophy*. This unclarity prevails since, to repeat, the pundit-style practice does not naturally lend itself to interactions with current systems of knowledge. The least that is expected here is that these consequences are debated with care and scholarship, and are not brushed aside because of alleged cultural inconvenience.

A vigorous pursuit of contemporary Western philosophy, then, is a perfectly legitimate practice in the Indian tradition even if this practice virtually *ignores* the classical Indian tradition. When we take even a cursory look at the actual body of current philosophical research in the West, we find that the literature hardly mentions authors of its *own* classical tradition. As expected in any developing discipline, this literature is primarily concerned with its contemporary authors, except of course when the focus is on the history of ideas. But this natural fact does not make Plato, Aristotle, Aquinas, Spinoza, and the like, fall out of the tradition. They are all there in the living but subliminal history of the discipline. We go back to them if the need arises, otherwise we just carry on with whatever problem currently occupies us as we stand on those great shoulders. What then is the argument, if any, that the practicing philosopher in India is failing in his professional task if he is unable to mention *Samkara, Nagarjuna, Prabhakara, Bhartrhari*, and other stalwarts of the past?

The argument for granting legitimate autonomy to the unhindered practice of contemporary Western philosophy *does not* prevent research programmes that attempt to directly link classical Indian philosophy with contemporary Western philosophy; it only casts doubt on the validity of the practice of Indian philosophy which is not professionally informed of contemporary Western philosophy. There are a number of possible points of contact that we will briefly look at in a moment. However, much caution is warranted for such research programmes if they are to be of any lasting value.

The central impediment for such projects, in philosophy rather than in history of philosophy, is the possible obsolescence of large areas of classical knowledge. It is one thing to admire the great edifice of thought that is enshrined in the texts; it is quite another to harness them for addressing current questions. In the lecture cited above, Amartya Sen rightly mentions the work of ancient scientists like *Āryabhatta, Varāhamihira, Bramhagupta* and others to illustrate the heterodox character of classical Indian thought. It does not follow that contemporary physicists and mathematicians are failing in their jobs if they are not directly engaged with these authors. For that matter, we do not expect Amartya Sen himself to build his economic theories principally on the basis of *Kautilya*'s work. Why should it be otherwise for philosophy in general? So, research programmes that attempt to link thoughts, which are widely separated in time, have a rather thin margin to play with.

Given this restriction, such interactions can take either of two forms. Although we have suggested a perspective in which contemporary Western philosophy is seen as a continuation of classical Indian philosophy, the strands of this continuity remain largely unexplored. So, the first form that an interactive research programme could take is to reconstruct the uncharted classical territory with tools of contemporary Western philosophy. In time, with sufficiently rich reconstructions in hand, the programme could be conducted even in the reverse direction: 'to interpret and critique some very fundamental concepts of Western thought in the language of Indian philosophy' (J.N. Mohanty, cited in Chakravarty, 1996, p. 9).

To be a bit more specific on how this programme might work, one may cite the curious fact that, for reasons that are just beginning to come under research, classical Indian philosophy at least since the *Mimamsakas* had been deeply concerned with questions regarding the nature and function of languages. Understanding the conditions of articulation was seen to be essential for understanding the conditions of valid knowledge (Mukherji 2000). Despite great internal differences, reflections on language dominated much of the debates in epistemology and metaphysics even when some proponents, such as the Buddhists, denied any significant role to language. In a general sense then, studies on language, thought, reality and knowledge, and their relations thereof, formed much of the content of Indian philosophy. To use a term popularized by Richard Rorty, this decisive *linguistic turn* took place in Western philosophy only around the turn of the century with the work of Gottlob

Frege, Bertrand Russell, Ludwig Wittgenstein, Edmund Husserl, and many others (Rorty 1967; Dummett 1992, 1993).

So, when we are seeking underlying connections between classical Indian philosophy and contemporary Western philosophy, it is no wonder that the more direct connections are likely to be found right here. Research on these possible connections is in its infancy. Yet a number of recent publications do seem to substantiate the point (Matilal and Shaw 1985; Zilberman 1988; Matilal 1990; Siderits 1991; Mohanty 1992). Some of J.F. Staal's work is also of related interest; in particular, Staal (1996). I have only listed some of the more prominent works of a general nature. See the references in these works for more focused scholarly work, as well as for some rather incomplete list of the original sources. Nevertheless, despite its philosophical interest, this form of research is likely to be historical in character for some time. This is because, before venturing to make any substantive philosophical comment, historical scholarship needs to settle on stable accounts of classical texts.

6.5 A Study of Needs

However, the first form of research just outlined might prepare the way for a second form of research which has direct philosophical significance. This form has not yet taken off the ground to my knowledge. So, I wish to spend some time on this. This form of research begins with some questions about the status of philosophy as it relates to other systems of knowledge. Given that the most profitable area of work is likely to be centered around topics such as language and knowledge as suggested above, philosophical inquiry must open itself, albeit critically, to vast developments in the adjacent sciences. By now, studies on language, cognition, consciousness, and the like are exploding areas of scientific research (Pinker 1995, 1997 for some popular review). Unless there is critical engagement between philosophy and these scientific disciplines, philosophy, *in either tradition*, is likely to acquire features of obsolescence.

Going by recent proclamations, the prospects aren't absurd at all (Mukherji 2003, *this volume*, Chap. 5). To take one quick example, Daniel Dennett (1995, p. 203) says, after a fascinating account of virus replication, that

an impersonal, unreflective, robotic, mindless little scrap of molecular machinery is the ultimate basis of all the agency, and hence meaning, and hence consciousness, in the universe.

Thus, to understand more about, say, agency, we need to understand more about 'molecular machinery'. Contemporary Western philosophy is already looking for ways to adjust itself to these new developments (Goldman 1985; Boden 1990; Hookway and Peterson 1993; Casati et al. 1994).

Although no research idea can be ruled out in advance, it seems unlikely that the study of 'molecular machinery' could be interestingly linked to classical Indian thought. Given the typical facts of underdevelopment, it is even more unlikely that these scientific studies could be pursued in India at their cutting edge. Roughly, then, the scenario is that since the study of 'molecular machinery' neither fits in with Indian philosophy nor can it be fruitfully conducted from here, meaningful philosophical practice is doomed in central areas such as epistemology, language, cognition, emotions, and the like. Philosophy shrinks to a 'spiritualistic' discipline, just as the revivalist, 'orientalist' view would want it to be.

So, the task is to confront Dennett's dangerous idea by examining the scope and the limits of current 'molecular machinery' , namely, current science. In my opinion, there are principled reasons that current science cannot advance beyond a point in these domains (Mukherji, 2006b, *this volume*, Chap. 11). The basic reason is that classical philosophical issues arise at a level of complexity that no 'molecular machinery' can hope to reach from what we can gather from the current character of science. This is not to rule out, of course, the possibility of what Chomsky (2000) has recently called *Ethnoscience*: systematic, perhaps 'naturalistic', inquiry into human common sense. Whether an envisaged ethnoscience will take the form of an account in terms of 'molecular machinery' is something we cannot even speculate upon for now.

Nevertheless, the present point is that, ethnoscience or not, we *do* have some tacit account of the world around us for us to lead our lives in the first place. Whether this account will stand the test of naturalistic scrutiny is an entirely different issue. Perhaps, the account we tacitly entertain is full of falsities. It is an account that we find personally and socially useful, nonetheless. The task of philosophy, in one sense of this catch-all nominal, then is to subject this account to critical reflection to see whether they can be—rather, what needs to be done such that these can be viewed as—rationally justified. To take a quick example, we tend to believe what we see from close quarters under normal lighting. Is that belief justified? What are the boundary conditions? What account of perceptual belief can we rationally furnish such that it turns out that it is not irrational to entertain such beliefs? Needless to say, issues become vastly more complex as we examine common entertainments of self and other minds, rules and obligations, durability of objects, repeatability of processes, and so on.

Now, it may turn out, as seems to be the case from recent findings, that formation of perceptual beliefs can be explained entirely on 'internal' grounds, that is, from properties of visual stimulation alone. In other words, there need not be an 'external' world for these beliefs to form (Ramachandran 2006). There is, then, a possible conflict between our common suppositions and the findings of naturalistic inquiry. Yet the common supposition itself is an overriding fact about the human condition. How do we justify it? The only course here is to seek some route to justification from within the network, so to speak, of common suppositions themselves; for example, suppositions regarding truth, validity, agreement, and action. In that sense, this form of justification—call it *philosophy*—demands a certain autonomy from naturalistic routes of justification, though the naturalistic route, *if* there is one, may be right about what the world is like.

It follows that classical philosophical issues in these domains reflect human *needs* rather than the 'order of things'. The study of human needs—what

Wittgenstein called 'philosophical clarification'—arises precisely at the limits of science, in the sense outlined. Study of concepts like knowledge, belief, meaning, truth, consciousness, etc. are not studies of properties of things such as mental/brain states; these are concepts that humans need and, therefore, construct, to carry on with their personal and social selves. If that is the case, then no development in science can overthrow *these* corners of philosophy. For example, it is sometimes suggested that a naturalistic inquiry into human mental life will not require the concept of belief (Stich 1983). Yet we cannot do without this concept in common life (Mukherji 2006a, *this volume*, Chap. 9). Needless to say, this study of needs requires to be done at each phase of scientific advancement to redraw the boundaries of philosophy.

The supposed autonomy of philosophy, however, has no historical privilege; it preserves its autonomy by constantly adjusting itself to other systems of knowledge, especially science. When this dialogue comes to a halt, the chances are that the falsities will now be promulgated even without an internal justification. A philosophy without engagement with other systems of knowledge is thus self-stultifying. Yet, the irony is that this very engagement of dialogue with other systems of knowledge might give rise to the illusion that philosophy is getting *submerged* in such systems. Uncritical openness *could* mean loss of autonomy. Ever since the advent of modern science, Western philosophy might have wavered uncertainly between these conflicting pulls of philosophy even though there has been a clear underlying distinction throughout.

One way to test the hypothesis just proposed is to examine the philosophical inquiry conducted in traditions where *modern* science didn't play any role at all, and where the study of needs—misleadingly called *ways of life*—had been explicitly advocated. That 'noise-free' environment will tell us how to strip away the scientific vestiges of Western philosophy, and to situate the rest for perennial human reflection (Mukherji 2006b). This is where the study of classical Indian philosophy takes centre stage. That philosophy itself must have interacted with other systems of knowledge, including forms of ancient Indian science, at certain stages of its development. It is doubtful, however, if the interactions were as frequent and thorough as in the case of modern Western philosophy. In any case, Indian philosophy stopped growing, as noted, before the entry of modern Western science in the Indian scene. So, the philosophy that the pundits preserved for centuries has remained untarnished—in a 'natural' state, so to speak—from the pulls just mentioned. We thus have an example of philosophy that retained a certain purity of form.

Here the fact that Indian philosophy stopped growing some centuries ago will not prominently affect the point of the enquiry, since it has been delinked from the factors which contribute to its apparent obsolescence. We will also learn how to continue to philosophize with those reflective tools once the barriers of era and mode of expression are carefully removed with scholarly study. Questions thus arise regarding whether the current academic structure of philosophy in India is prepared to face the demanding tasks sketched above. An examination of such questions however is a topic for another essay.

References

Casati, R., B. Smith, and G. White (Eds.). 1994. *Philosophy and the Cognitive Sciences*. Vienna: Holder-Pichler-Temsky.

Chakravarty, A. 1996. The absence of a philosopher. *Epistemology, Meaning and Metaphysics after Matilal: Studies in Humanities and Social Sciences*, III(2, Winter), 1–11.

Chomsky, N. 2000. *New Horizons in the Study of Language and Mind*. Cambridge: Cambridge University Press.

Boden, M. (Ed.). 1990. *The Philosophy of Artificial Intelligence*. Oxford: Oxford University Press.

Dennett, D.C. 1995. *Darwin's Dangerous Idea*. London: Penguin Books.

Dummett, M. 1992. *Origins of Analytic Philosophy*. Oxford: Clarendon Press.

Dummett, M. 1993. *The Seas of Language*. Oxford: Clarendon Press.

Dummett, M. 1996. Motilal's mission: a memorial address. *Epistemology, Meaning and Metaphysics after Matilal: Studies in Humanities and Social Sciences*, III(2, Winter), 13–17.

Goldman, A. 1985. *Epistemology and Cognition*. Cambridge: Harvard University Press.

Hookway, C., and D. Peterson (Eds.). 1993. *Philosophy and Cognitive Science*. Cambridge: Cambridge University Press.

Matilal, B.K. 1990. *The Word and the World*. Delhi: Motilal Banarasidas.

Matilal, B.K., and J.L. Shaw. 1985. *Analytical Philosophy in Comparative Perspective: Exploratory Essays in Current Theories and Classical Indian Theories of Meaning and Reference*. Dordrecht: D. Reidel.

Mohanty, J.N. 1992. *Reason and Tradition in Indian Thought*. New Delhi: Oxford University Press.

Mukherji, N. 1996. Realms of Satchidananda Murty. *Summerhill Review*, Shimla, August.

Mukherji, N. 1997. Classical Indian theories of language and contemporary syntax research. *Summerhill Review*, Shimla, November.

Mukherji, N. 2000. Traditions and concept of knowledge. In *Science and Tradition*, A. Raina, B. N. Patnaik, and M. Chadha (Eds.), 14–37. Indian Institute of Advanced Study: Shimla.

Mukherji, N. 2002. Academic philosophy in India. *Economic and Political Weekly* 37(10), March.

Mukherji, N. 2003. Skeptical politics. In *Truth and Value: Essays in Honour of Pabitra Kumar Roy*, R. Ghosh (Rd.). New Delhi: New Bharatiya Book Corporation.

Mukherji, N. 2006a. Beliefs and believers. *Journal of Philosophy*. Calcutta University, December.

Mukherji, Nirmalangshu. 2006b. Textuality and common life. In *Literature and Philosophy*, S. Chaudhury (Ed.). Papyrus: Kolkata.

Pinker, S. 1995. *The Language Instinct*. New York: Harper-Collins.

Pinker, S. 1997. *How The Mind Works*. New York: Harper-Collins.

Popper, K. 1962. *Open Society and its Enemies*. London: Routledge and Kegan Paul.

Ramachandran, V.S. 2006. *Phantoms in the Brain: Human Nature and the Architecture of the Mind*. New York: Harper-Collins Publishers.

Rorty, R. (ed.). 1967. *The Linguistic Turn*. Chicago: University of Chicago Press.

Sen, A. 2001. History and the enterprise of knowledge. *Frontline*, February 2, 86–91.

Sen, A. 2005. *The Argumentative Indian*. London: Allen Lane.

Siderits, M. 1991. *Indian Philosophy of Language: Studies in Selected Issues*. Dordrecht: Kluwer Academic Publishers.

Staal, J.F. 1996. *Rituals and Mantras*. Delhi: Motilal Banarasidas.

Stich, S. 1983. *From Folk Psychology to Cognitive Science: The Case Against Beliefs*. Cambridge: MIT Press.

Zilberman, D.B. 1988. *The Birth of Meaning in Hindu Thought*. Dordrecht: D. Reidel.

Chapter 7
Yearning for Consciousness

> *Einstein's brain turned out to be no bigger than normal. Just as Einstein captured the essence of energy and matter in his famous equation, so we seek to capture the essence of genius. Our pursuit perhaps reveals more about ourselves than about the nature of genius.*
>
> Hao Wang (1987)

In this chapter, I will suggest that the concept of consciousness, as commonly envisaged, is something we need even if nothing in the world falls under it. In subsequent chapters, I will show that similar remarks apply to the concepts of belief and knowledge. In the context of contemporary philosophy, the suggested shift in the form of inquiry arises as follows.

The discipline of *philosophy of mind*, as the name suggests, may be viewed as a conceptual investigation of the mind in terms of its 'mentalistic' aspects. These mentalistic aspects include consciousness, perception, knowledge, belief and intentionality. Sometimes some of these concepts are examined in terms of analysis of the linguistic contexts which appear to exhibit the need for these concepts. For example, putative mental states like belief and knowledge are examined via what have come to be known as 'propositional attitudes' like *Galileo believed that the Earth is round*. It is thought that a detailed understanding of what such linguistic expressions mean in their standard contexts of use will throw light on the character of the corresponding mental states. With the focus on language, the study of mind becomes species-specific, as desired in the Cartesian angle on this topic (Mukherji 2000a; see also *this volume*, Chap. 3). Also, since the proposed philosophical study is focused on ordinary, daily uses of the listed mentalistic concepts, it might appear to throw light on the common, universal—*folk psychological*—aspects of human nature.

However, such linguistic approaches do not exhaust the philosophical study of mind. Even though consciousness as a mental state is often studied via analysis of (first-person) reports of experience such as *I am in pain*, the concept is seldom viewed as language-related. Sometimes mental concepts are studied more directly through introspection and analysis of specific states of experience and behaviour. In

© Springer Nature Singapore Pte Ltd 2017
N. Mukherji, *Reflections on Human Inquiry*,
DOI 10.1007/978-981-10-5364-1_7

recent decades, some of these concepts—especially the concept of consciousness—have been vigorously studied via experimental investigations on the brain. These approaches appear to hold the promise of a genuine science of the mind.

Despite these appearances, my contention is that such philosophical and scientific studies, focused on ordinary mentalistic concepts, cannot really be viewed as a naturalistic study of the mind. I will suggest that a biologically-anchored, theoretical concept of mind requires the mind to be a (genuine) property of every individual mind/brain of a species—often known as the requirement of methodological solipsism (Fodor 1980). In contrast, the philosophical concepts of consciousness, belief and knowledge are primarily interpersonal social devices; these are more likely to be normative concepts for that reason, on a par with concepts used in inquiry of values such as ethics and aesthetics, rather than in naturalistic inquiry like physics and biology.

7.1 The Sentient Subject

The study of the thinking, sentient human subject has always been a central concern in philosophy in any tradition. However, only with the Cartesian rationalist tradition did the concern directly relate to the concept of mind as a separate substance, an additional joint of nature. As the Cartesian tradition of substance dualism lost its appeal in subsequent centuries due to serious challenges to the idea of a separate substance by empiricists like David Hume and John Locke, the discussion of mind itself, as a substantive concept, was progressively abandoned. In contemporary times, the situation for the Cartesian tradition worsened even further after Gilbert Ryle's influential critique of the concept of mind as the ghost in the machine (Ryle 1949).

Nonetheless, the concern about the sentient thinking subject remained, especially after the work of Immanuel Kant, because the subject was viewed as the centre of the complex network that related language, thought and reality: the domain of human knowledge. Thinking of human belief as a fact about humans, it is natural to view human belief as the content of mental states, states that humans attain when they have belief. Beliefs are thus viewed as contentful mental states *par excellence* of a subject. The step from belief to knowledge is deemed natural since knowledge is taken to be a species of belief: knowledge signals the attainment of a restricted kind of belief-state, namely, a state of true belief for which the subject has evidence.

It was then thought that specific beliefs can be identified in terms of structured meaningful propositions such as *that the Earth is round*. The proposition, a linguistic entity, represents the belief that the Earth is round by a systematic grammatical construction out of individual meanings of words such as *Earth* and *round*. These sounds are phenomena in the external world to be accessed by perceptual systems, but the meanings of these words must themselves be mental entities since they constitute the mental states of the typically sentient subject: the content of the belief that the Earth is round is constituted of mental entities "Earth" and "round";

these entities endow the sounds *Earth* and *round* with meaning. We thus have a set of mentalistic concepts: belief, knowledge, meaning, consciousness. The 'Cartesian' angle on these concepts is hard to miss.

Since the concepts of belief and knowledge in their ordinary usage are taken to denote mental states, some notion of mind is at least indirectly implicated, although a direct mention of it is forbidden due to Rylean strictures. In this indirect sense, concepts of belief and knowledge define the contours of contemporary philosophy of mind, and some of cognitive science, that takes the form of (study of) folk psychology. I will discuss the concepts of belief and knowledge more fully in the next two chapters.

Similar remarks apply to the concept of consciousness. It is a ubiquitous part of folk psychology that we view human subjects as beings which routinely attain states of consciousness. For example, John Searle (1992, p. xii) declares at the very beginning of his study of the mind that we 'all have inner subjective qualitative states of consciousness, and we have intrinsically intentional mental states such as beliefs and desires, intentions and perceptions'. Colin McGinn (1989) opens his influential paper on the mind–body problem with the observation that philosophers have been trying for a long time to solve the specific problem of consciousness which continues to be 'the hard nut of the mind—body problem'. We need some explanation of why we have these mentalistic concepts—belief, knowledge, meaning and consciousness—and what they do for us. The study of these concepts then qualifies as a study of the mind, in the indirect sense outlined.

Cutting through many-dimensional controversies covered in a vast literature, two broad perspectives have emerged in the philosophical literature. These perspectives are in serious conflict. The first perspective, often called *folk psychology* as noted, says that the availability of these concepts in fact points towards an implicit and largely correct theory of mind; the task is to make it explicit. In their ordinary usage, these concepts are already laden with explanatory value; all we need is to make proper scientific use of them. The neuroscientific perspective says, on the other hand, that these concepts have a value at most as components of a false theory; a genuine theory of cognition will dispense with these concepts. Each of the ordinary concepts of belief (Stitch 1983) and consciousness (Dennett 1991a, b) are to be eliminated from a putative science of the mind (Churchland and Sejnowsky 1992).

The debate has been stultifying because, while folk psychology is untenable at various points (Mukherji 2006; *this volume*, Chapter), neuroscience has done nothing to replace it in the critical cases (Mukherji 1990; Bennett and Hacker 2003). In particular, as we will see, fatal problems of explanation seem to block any coherent neuroscientific account of consciousness, not to mention beliefs and meanings. Although a careful review of the (astronomical) literature is warranted at this point—a task that is beyond the scope of this chapter—it is not entirely unfair to surmise that neither folk psychology nor neuroscience has much explanatory chance for the mental domain as commonly envisaged. Suppose so. Nevertheless, these concepts are here to stay with us in ordinary discourse. What do they do for us?

What I propose to do is to move away from the folk psychology/neuroscience debate, and launch an independent philosophical examination of these concepts to see if a general account of their value can be extracted *outside* the theory of mind. One option that seems particularly promising in the given historical scenario is that these concepts are not designed to play (genuine) explanatory roles in naturalistic theories at all; their value appears to be located elsewhere. As noted, I will concentrate on the concept of consciousness for the rest of this chapter; concepts of belief and knowledge will be taken up in the next two chapters.

The strategy adopted here is different from other attempts to set the problem of consciousness aside. For example, in the paper cited earlier, McGinn (1989) suggests that the problem of consciousness falls under what he calls the principle of *Cognitive Closure*: 'A type of mind M is cognitively closed with respect to a property P (or theory T) if and only if the concept-forming procedures at T's disposal cannot extend to a grasp of P (or an understanding of T).' McGinn thinks that consciousness is a property that we are cognitively incapable of understanding; in that sense, consciousness is not a problem, but a *mystery*.

Following Thomas Nagel and others, I have myself suggested a more general version of the closure principle earlier to raise the design problem: our grasp of the world is restricted to what kind of creature we are (Mukherji 2010, Chap. 1; also, *this volume*, Chap. 2). However, I do not think that consciousness is a mystery that cannot be understood by the human mind; the widespread misconception of consciousness as a mental phenomenon promotes the alleged mystery. As proposed, consciousness can well be understood as a normative/ascriptive concept; to view consciousness as a real/naturalistic property of human nature is probably not warranted. However, several qualifications are needed to properly articulate the suggestion, as we will see.

7.2   Problem of Explanatory Gap

The general difficulty with the problem of consciousness—unlike the very similar problem of God as we will see—is that its solution seems, at once, to be urgent and elusive. In other words, it seems that we cannot do without the concept of consciousness while, despite voluminous discussion especially in recent years, the concept continues to resist elucidation. Not only that we do not seem to have made much progress with the problem, we find it difficult to form an idea of what a progress in understanding is supposed to look like, as McGinn (1989) and many others have pointed out. It is this apparently 'unsolvable' aspect of the problem that interests me in this chapter.

The urgency of the problem of consciousness arises from the fact that, as John Searle observes, we 'all have inner subjective qualitative states of consciousness'. To bring out the ubiquitous character of consciousness, David Chalmers (1996, p. xii) introduces his influential work on the conscious mind as follows:

> I find myself absorbed in an orange sensation, and *something is going on*. There is something that needs explaining, even after we have explained the process of discrimination and action: there is the experience.

It is interesting that Chalmers reports a phenomenon in the first person. It is a report where the phenomenon has to do with (someone) having or undergoing an orange sensation. We will soon examine what it all means. However, we may note at once that an experience pertains to an individual sentient being. My (current) experience of the computer screen is entirely confined to the working of *my* cognitive faculties during this stretch; no one else can have *this* experience, even if someone is sitting next to me right now staring at the same screen; my best guess is that that person is undergoing his *own* experience. Therefore, strictly speaking, an experience cannot be shared, forcing Chalmers to shift to the first person. In this sense, experiences are subjective and qualitative, as Searle observes.

Since experiences are both subjective and ubiquitous parts of our lives, their occurrences call for explanation. The question arises as to what can legitimately be the form of such as explanation. Since, by the nature of the case, there is no third-person description of the phenomenon itself—that is, I cannot describe what is it for Chalmers to undergo the reported sensation, I can only describe mine—all we can do is to look for the unique conditions that are 'objectively' satisfied at the locus of the concerned sensation. Setting a host of conceptual issues aside, suppose we agree that for someone S to have the sensation X, S's brain has to attain a certain state Y such that the attainment of Y is uniquely correlated with X, *without residue*. Under those circumstances—call it, *mind–brain identity*—it should be legitimate to say that once we have given a complete third-person description of Y, we have shown what is it for the world to contain X; there is nothing more to show.

So, what is it that we need to show? Here Chalmers says that it is not enough that we have explained the 'process of discrimination and action', we need to explain the (qualitative, subjective) experience itself. For example, Ned Block (2007) reports interesting work by Nancy Kanwisher and colleagues who showed that there is a strong correlation between face-experiences and the activation in a very specific area of the brain located at the bottom of the temporal lobe in the right hemisphere, called the *fusiform face area*. Block views the fusiform face area as an informationally encapsulated Fodorian module (Fodor 1983), a view that raises problems for the reportability of these experiences according to Block; I set such problems aside. Suppose there are other 'modules' for experience of fruits, canines, fuzzy drinks, etc. The working of these modules, then, would count as (a) discriminating various stimulus items, if any, and (b) constituting the neural actions that lead to these discriminating representations.

Still, a description of a module in a particular state does not amount to a description of the resulting experience—*what it feels like*—of faces, fruits, fuzzy drinks and the like. That is the problem raised by Chalmers: there is a crucial residue. Such advances in neurosciences thus fail to answer Julian Huxley's classic question (cited in McGinn 1989),

> How it is that anything so remarkable as a state of consciousness comes about as a result of initiating nerve tissue, is just as unaccountable as the appearance of the Djin, where Aladdin rubbed his lamp in the story.

In a later paper, Block (2009) restates Huxley's problem by observing that 'we have no idea why the neural basis of an experience is the neural basis of that experience rather than another experience or no experience at all'. Block calls this the *problem of explanatory gap.*

Having said so, Block proceeds to examine a variety of options proposed in the current scientific literature to weigh their merits: biological state of the brain, global workspaces, and hierarchy of integrated physical systems, and several sub-versions thereof. Block concludes that the problem of explanatory gap is best addressed in terms of the biological state of the brain; the details of Block's argument are beyond the scope of this chapter. As far as I could follow, Block's maximum contention is that, insofar as the problem of explanatory gap is concerned, the biological approach is superior to the other approaches because the biological state of the brain 'matters'; Block did not claim that the problem has been sufficiently understood in the biological approach, not to mention being solved. Let us assume that, despite Block's preference, the state of the art is such that we have no clue as yet to the real 'hard nut' of the problem of explanatory gap: the problem of residue, that *something is going on.*

But suppose that, contrary to the state of the art, some detailed account of the activation of the brain does furnish a satisfactory account of the feel of what it is like to experience the computer screen; that is, however incredible it may sound, suppose that the biological account actually displays the condition of unique correlation between some state of a given brain and the subject's reliable report, in the first person, of a particular experience. Will that count as an *account* of phenomenal consciousness, even if we have given up any form of dualism to agree that the brain is the seat of consciousness if anything is? Is the brain the right object to which the concept of consciousness legitimately applies? Is the brain, at that unique moment, undergoing phenomenal consciousness?

This problem can be raised in delineable steps. Recall the discovery of the fusiform face area that gets activated when faces appear in the visual scene. After describing the module, Block (2009, p. 1112) points out at once that the existence of the dedicated area does not mean that the area by itself gives rise to the experience of faces: 'No one thinks that a section of visual cortex in a bottle would be conscious.' So, Block emphasizes that it is not just the fusiform area but the total neural system with the entire brain that is involved. Thus, we generalize the application of consciousness from a dedicated area to the brain itself. But even that seems to be insufficient.

Tonini and Koch (2015) point out that a unique correlation between subjective states and states of the brain will fail to apply to 'patients with a few remaining islands of functioning cortex, pre-term infants, non-mammalian species, and machines that are rapidly outperforming people at driving, recognizing faces and objects, and answering difficult questions'. I am not suggesting that the concept of consciousness does legitimately apply to each of these cases, but the point is that we cannot even investigate the issue if certain brain states are held to be necessary for attaining conscious states. So, we need to generalize beyond brains. In any case, Block maintains silence on whether a (full) brain in a vat would be deemed conscious, even if Block is convinced that (just) the visual cortex in a bottle wouldn't be conscious.

Concerning the old issue of whether computers think, Noam Chomsky replied that legs don't walk, people do, even if people walk with legs; similarly, computers or brains don't think, people do. The trouble with neural correlationism is that it simply misses the grain of explanation that involves the entire organism to which the concept of consciousness typically applies. The objection is bolstered by the fact that the common notion of consciousness, which is the only notion currently at issue, does not refer to states of brains at all.

It is not at all implausible to think of people correctly applying the notion of consciousness in a variety of circumstances without any knowledge about under-lying brains; otherwise, most fables will not work—the frog simply didn't have the brain of the prince. Of course, neuroscientists are free to use technical terms to denote the relevant unique activation states, if any, of the brain, which they believe instantiate conscious states of a subject. But that nomenclature will apply to the subject's brain, not to the subject herself. Having noted the crucial distinction, we may as well hold on to biological correlation as the only physical basis of consciousness.

7.3 What Is Experience?

After criticizing the explanatory role of biological/neural correlationism, Tonini and Koch (2015) propose their own theory of consciousness that aims to clarify afresh 'what experience is and what type of physical systems can have it'. To that end, they present what they call *Integrated Information Theory*, details of which we set aside. Tonini and Koch claim that their theory successfully answers questions about nonhuman animals, people in coma, digital computers, etc.; for example, they ascribe consciousness to nonhuman animals, even 'some very simple ones', but not to digital computers.

Block (2009) views Integrated Information Theory as a species of functionalism 'according to which consciousness is characterized by an abstract structure that does not include the messy details of neuroscience'. Block complains that Integrated Information Theory fails to distinguish between having intelligence and having consciousness when, according to Block, 'separation of consciousness and cognition has been crucial to the success of the scientific study of consciousness'. Setting aside Block's view of Integrated Information Theory, the present point of interest is that, in order to argue for the separation of consciousness and cognition, Block also contends that 'mice or even lower animals might have phenomenal consciousness'. It appears that, independently of whether they adopt functionalist or biological theories of consciousness, researchers agree on the application of the concept of consciousness to nonhuman animals, perhaps going down to 'lower' and 'simple' life-forms.

On the face of it, such a large agreement on lower animals having phenomenal consciousness—*along with humans*—carries much physiological basis. Phenomenal consciousness seems to be a property of organisms that marshal some or other sensory system. As some stimulus triggers the sensory system under appropriate 'non-anesthetic' conditions, the organism attains a state of phenomenal consciousness as it undergoes the stimulus-experience. So, if the organism is endowed with receptors that detect specific colours such as orange, the organism will *undergo* an orange sensation, just as David Chalmers did.

It could be that the occurrence of sensations require certain developed forms of sensory apparatus such that very low forms like algae and molluscs are excluded. But it is very implausible indeed to restrict the phenomenon of experience only to human brains and the accompanying nervous system. Let us grant, therefore, that most organisms, by virtue of *being* an organism, have or undergo experiences; when they do so, they are in a state of phenomenal consciousness. Needless to say, all the accompanying problems of the first person and explanatory gap continue as before. We will never know introspectively what it feels like to be a mouse or a desert insect even if, incredibly, we have fully charted out *that* unique state of the insect's brain. But then, as noted, if we do have a unique chart, it is unclear what else there is to know.

In order to bring out this obvious point about animal consciousness, I did not marshal any sophisticated facts about brains and neural connections; a pedestrian view of brain-states was enough. As far as I know, the Seventeenth-century French philosopher René Descartes was well acquainted with whatever was then known about the anatomy and physiology of brains, much more than I ever did. Yet, according to standard interpretations, Descartes insisted that only humans are conscious; all other organisms are mere automata, machines that push and pull but don't feel anything. Thus, Peter Singer (1976, p. 217) writes:

> In the philosophy of Descartes the Christian doctrine that animals do not have immortal souls has the extraordinary consequence that they do not have consciousness either. They are mere machines, automata. They experience neither pleasure nor pain, nor anything else.

Apart from sounding deeply unethical, Descartes' view, as stated by Singer, seems obviously false since, as noted, it is implausible to deny that animals undergo experiences and, thus, attain states of phenomenal consciousness.

But what did Descartes in fact say and mean? In his famous letter to Henry More, dated 5 February 1649 (Cottingham et al. 1991, p. 366), Descartes categorically asserted that 'I do not deny life to animals... and I do not even deny sensation, in so far as it depends on a bodily organ.' In effect, Descartes readily agreed with the biological correlationism advanced by Ned Block and many other authors. Moreover, it is well known that Descartes denied that animals have thoughts. Thus, Descartes kept cognition (=thoughts) and consciousness (=sensations) strictly separate for animals, as recommended by Block and others. However, Descartes did not keep cognition and consciousness separate *for humans*; earlier in the same letter, Descartes stated that 'thought is included in our mode of sensation' (p. 365).

Setting aside Descartes' alleged elaboration of this human-specific position in terms of 'Christian doctrines', let us ask: what exactly was the phenomenon David Chalmers asked an explanation for? *What* was going on? There was obviously an orange sensation going on. If that was all there was to the phenomenon, then (unique) biological correlations is certainly a fine explanation when available. But Chalmers asked for more than such 'discrimination and action', as we saw. Why? Because he found himself 'absorbed in an orange sensation'. Chalmers was not just undergoing an orange sensation, looked at from the outside as it were, he *found himself absorbed* in the sensation from the inside. That's what Chalmers, rightly in my view, calls an *experience*.

Do animals, to whom Descartes cheerfully assigned the property of undergoing sensations, also *find* themselves absorbed in those sensations in the sense in which Chalmers finds himself absorbed? Descartes would have answered in the negative since, for him, animal mode of sensation does not include thought, the thought *that I am having orange sensation*. We know that humans do because humans *say so*. It is for this reason that—otherwise abruptly—Descartes shifted to a quick discussion of lack of speech in animals in the same letter to Henry More: 'speech is the only certain sign of thought hidden in a body' (p. 366). Several centuries later, Donald Davidson (1975) reached the same conclusion with more sophisticated arguments: animals don't have thoughts because they don't talk.

It could be that both Descartes and Davidson are using unnecessarily narrow conception of talk and speech to exclude the animals. Perhaps there are gentler notions of (structured) thought that might favourably apply to some animals of sufficient neural complexity. Psychologists often conduct behavioural experiments on animals to elicit structured, discriminatory responses. For example, some authors suggest that monkeys may be viewed as 'reliable reporters of their own visual experience' (Flanagan, 1998, p. 143). Flanagan was inspired to form this uplifting opinion about monkeys because monkeys have been trained to press a bar depending on whether they see a downward or an upward motion. Then, using the phenomenon of binocular rivalry, downward motion is presented to one eye, while upward motion is presented to the other simultaneously. Apparently, monkeys 'report' just as humans do. I set aside the issue of whether this example is sufficient to count as structured thought, or whether it is just (stimulus-dependent) bodily response as Descartes would have insisted. The issue is predominantly empirical.

The basic issue is not whether the notions of experience and phenomenal consciousness legitimately apply to nonhuman animals. The Cartesian point is that these notions legitimately apply to thoughtful modes of sensation. Since a sensation, by definition, pertains to individual organisms, a thought of *that* sensation can only be a first-person thought marked by the use of the pronoun *I* or its equivalent. This roundabout way of bringing out the Cartesian point basically means that an organism needs to have the *concept* of sensation, experience, consciousness and the like to *find* that something is going on; otherwise, something just goes on. Strictly speaking, undergoing an orange sensation is not an experience, finding oneself so undergoing *is*. Obviously, what one *finds* is not the heightened state of the brain, but the resulting *feel*.

There are various philosophical theories that introduce an additional level of awareness to capture the point: one is not only aware of or having the orange sensation, but also aware that one is so aware. Apart from other philosophical difficulties, the move is plainly unwarranted. There are not two awarenesses: one for the orange sensation, and the other for the feel. Metaphysically speaking, there is exactly one event, namely, the unique state of the brain under the assumption of biological correlationism. To feel is to *be* in that neural state, period. As Descartes puts it, there is only one experience in which thought is *included*. It is included for linguistic creatures for whom experiences are typically encoded in structured thoughts. This happens even in those cases where the 'discrimination' available to the subject of experience is merely *that there is an experience* when the subject has no clue about what the current experience is an experience of. This much should be enough to include cases, if any, of so-called 'nonconceptual perception'. For Descartes, even 'bare' experiences take the form of structured thoughts.

Once linguistic creatures carry the burden of structured thoughts as a biological endowment, the question arises as to the character of the basic structure of thoughts. It is generally agreed that thoughts are couched under the propositional form *that A is B* which has marked constituents that include stuff like nominals and verbals. Setting much controversy and qualifications aside, nominals and verbals carry meanings which basically inform what they are about. So the thought "that I am having an orange sensation" includes the complex nominal *orange sensation* as a constituent; that is the linguistic burden. What, then, is *orange sensation* about? It can't be about the unique neural state of the body to which the subject typically does not have any access while undergoing the sensation. What the subject appears to have access to is the 'feel'. But metaphysically speaking, the feel is just the state of the brain embedded in the relevant neural system, as noted. It is a reasonable conclusion that the feel is a necessary fiction by dint of the thoughtful mode of undergoing sensations. There is, of course, the discriminatory experience of something orange; that's what the visual system is for. But I doubt that it is generally the case that the experience of something orange is accompanied by a 'feel'.

Is the feel entirely a fiction? I am unsure what the answer is. I just said that the subject does not have introspective access to the state of the biological system while undergoing a specific orange sensation. That seems to be the *typical* case. But it does not rule out the possibility that, on occasion, subjects may even have intro- spective access to the brain itself. We do have introspective access to states of other organs such as heart and stomach, not to mention the obvious case of genitals. Those experiences are, of course, 'guided' by the neural system. Are there similar experiences of the brain itself?

Anecdotally, I can report that, on several occasions during particular phases of high fever, it appears as if the brain itself is the object of experience. While one is still fully awake, one is unable to focus on any specific object of thought through standard perceptual means due to high fever; in fact, it is difficult to keep one's eyes open. Yet there seems to be the experience of a dark void pulsating in the head; it is very different from even crushing headaches felt in localized areas. Supposing this to be a genuine 'objectual' experience, it will not be surprising if it always sub-liminally accompanies specific 'external' sensations. Even then the point remains that we simply do not have introspective access to the specific state of the brain *undergoing* orange sensation; so, the feel of orange sensation is most likely to be a fiction in the linguistic mode as suggested.

The shift of talk, as above, from visual experience to painful experience could be an indicator of how the fiction arises. We genuinely report feels when in pain because we have at least partial introspective access to the state of the body. In those cases, the conception of the feel could be viewed as a response of the body itself to some (injured) state of the body. That is why when we are in (physical) pain, we visit the doctor, not the psychologist, without taking a stand on dualism. However, talk of feel when looking at the computer screen sounds a bit odd, unless the screen is glowing or something such that it hurts. Yet it cannot be denied that my visual experience of the computer screen is embedded in the thought that I am having a visual experience of a computer screen. In this case, the thought consti-tutes of the referent of *visual experience* where the referent is typically neither some introspection on the state of the body nor some visual image; needless to say, the computer screen is not the referent of *visual experience*, it is the referent of *computer screen*. My contention is that in such cases the ascription of some feel/sensation appears to be a bit of a fiction.

To sum up, the concept of phenomenal consciousness seems to have a variety of conflicting pulls. In some form or other, phenomenal consciousness is inevitable for organisms with developed sensory systems. The first problem is with its inherently first-person character. It arises because the occurrences of experiences are subjec-tive in the sense outlined. When we try to overcome this problem with, say, a (third-person) biological story that correlates a unique state of the brain with occasions for subjective experiential states, an explanatory gap shows up. Since the problem sounds more empirical than conceptual, suppose the gap is somehow bridged in some future science, perhaps in the form of a complete description of the underlying biological system.

Even then it appears that the purported biological explanation will have the wrong grain because the first-person states are the states of a subject as a person, rather than the states of the subject's brain. Keeping exclusively to subjects' reports, it is also extremely unclear what the report of experiences are *about* once we have delinked those reports from the unique brain-states correlated with those experi-ences. In fact, in some cases, it would seem that the reports are about nothing, the 'feel' involved in experiences could be fictions. Yet, and this is the near-fatal point, the *talk* of experiences just cannot be given up or eliminated because the talk simply reflects the structured thought that is *included* in subjective experiences, at least in the human case.

Humans seem to be condemned to entertain the concept of consciousness as part of their cognitive design, very much like they are condemned to entertain the concept of an (external) reality (see *this volume*, Chap. 2). That is all I have to say for now on this elusive topic; I am aware that it is difficult to say something definite on the metaphysical aspect of phenomenal consciousness. Be that as it may with the state of scientific explanation of consciousness, is there some other way of explaining the talk of phenomenal consciousness in common life?

7.4   Ascriptions and Descriptions

I wish to argue for a framework in which the general impasse just sketched may be resolved or, at least, set aside. In particular, I will attempt to show that the concept of consciousness could be one of those whose importance and use in our common life need not suggest that it pick out some property of some delineable object. In that the concept is not too dissimilar to the concept of God. So it is not very surprising that the concepts of consciousness and God—or, of a supreme reality in the form of pure cosmic consciousness—are often held to go together in classical theological literature. In fact, subliminal appeal to some notion of 'supreme reality' as the fundamental aspect of the universe may be detected even in some formidable scientific literature on consciousness (Penrose 1994; Chalmers 1996). If the history of the discussion of the concept of God is any guide, it is perhaps futile to look for the properties of some entity designated by the concept; the (human) interest—Wang's Puzzle—is to ask why people commonly and universally entertain the concept at all despite the futility of the ontological search. What needs does the concept satisfy?

In the sense under discussion here, we *ascribe* a concept to an object solely to display our, for want of a better term, *personal* attitudes towards the objects which fall in the domain of application of the concept. What attitude we hold towards these objects depends partly on our rational, reflective *needs*, rather than on the internal/intrinsic properties of these objects. In this, ascriptions differ sharply from *descriptions* in which we use some concepts with an intention to pick out genuine properties of objects to which we apply these concepts; whether we do in fact pick out genuine properties is a wholly different matter.[1] One hallmark of the distinction between ascriptions and descriptions is that, in ascribing a concept to an object, we cheerfully entertain the possibility that someone might withhold the said ascription or ascribe an opposite concept, and that we may *both* be right. Obviously, we do not describe something with the expectation that someone might give an opposite description, and we both come out right; if that puzzling situation does arise, we look for a new concept, a new description. Some examples are in order at this point.

When we call a painting *beautiful* we do so because we like it, or enjoy watching it, or are fascinated by it, or whatever; we do not think that beauty, unlike the coats of paint or the size of the paper, is an intrinsic property of the object so admired.[2]

---

[1]Before we proceed, let me emphasize that it is *need*, rather than *interest*. So ascriptivism cannot be rephrased as interest-relative worldviews (Goodman 1978; Putnam 1981).

[2]Although the example might suggest otherwise, the distinction between primary and secondary qualities is not relevant here; there are other related distinctions which we need to set aside.

Therefore, it is not puzzling, though annoying, that someone else might call it *ugly* or *uninteresting*. This analysis persists even if we say *That piece of work contains beauty* or *That piece of music has great beauty*. But when we say *Photons have no rest mass* or that *Electrons are spin-half particles*, we literally mean so. A supporting test of this point is that we say *Electrons have the property of spin-half*, but we cannot say *Guernica has the property of great beauty*. So one way of marking the distinction between descriptions and ascriptions is in terms of where *property*-talk applies and where it does not apply; as noted, *has*-talk or *is*-talk or *contains*-talk are not definite guides to the distinction.

I must hasten to note that this analysis does not render the concept of beauty to be superfluous. We need the concept to characterize an artistic object, and this characterization is not available without the concepts of beauty, harmony, elegance and the like. By characterizing an object as an artistic one we form a favourable evaluative attitude towards that object; that is the need which the concept of beauty anchors. It appears that this form of analysis extends to moral concepts such as honest, truthful or coward; given that the uses of these concepts are geared solely to display our evaluative attitudes towards persons, we need not look for cowardice-genes or honesty-mutations.

However, it is not ruled out that some of the concepts, or parts of them, which are marked by their ascriptive use, may in fact turn out to determine genuine properties of things. Consider the concept of schizophrenia. As commonly used, it displays our attitude of puzzlement towards certain persons. Earlier textbooks of psychology simply picked up this ascriptive common usage, and used the concept to describe certain psychotic conditions. It turns out, however, that certain restricted forms of schizophrenia, some of which *do not* fall under the common usage enshrined in the textbooks, may in fact have a neural or even a genetic basis. On the other hand, several conditions commonly 'described' as schizophrenia have no such basis, and can be traced to faulty and jaundiced evaluation of behaviour. I return to this concept again.

As a historical aside, some of what I said above may be usefully compared and contrasted with the work of Peter Strawson (1957). In a somewhat similar vein, Strawson conceived of a discipline called *descriptive metaphysics* which is geared to the study of a 'massive ahistorical core' of concepts which ought to be available for us to understand our relations with the world and with one another. These concepts are 'ahistorical' in the sense that their use is not affected by the progress of science whereby some of these concepts may turn out not to designate genuine features of the universe at all; whether or not they are genuine, we ought to have them since we *need* them to form our basic (communicable) understanding of the world *before* we embark on the course of science.

As examples, Strawson lists the following concepts: the concept of a single space-time continuum, the concepts denoted by some M(aterial) predicates and some P(ersonal) predicates. We need not ask how these concepts originate; we just ask what role they play in our understanding given that they somehow originate. Strawson's work describes the link and the hierarchy among these concepts. However, it is unclear what status these concepts have in the *total* body of concepts

—scientific and otherwise—we come to have. What is the nature of the task that concerns the study of *these* concepts as opposed to the (scientific) task of uncovering genuine properties of things? As a clue to the problem, Strawson suggests elsewhere (Strawson 1992, p. 7) that

> the philosopher labours to produce a systematic account of the general *conceptual structure* of which our *daily practice* shows us to have a tacit and unconscious mastery. (emphasis added)

In other words, the philosopher aims at finding the 'theory of employment' of this conceptual equipment we unconsciously use. The philosophical 'theory' then locates the efficacy of these concepts in our coming to grips with ourselves without advocating that these concepts in fact display features of the world; in that restricted sense, these concepts highlight features of ourselves rather than features of how things are. Strawson's project then is better titled as *Metaphysics of Ascriptions*.[3]

As an illustration of Strawson's theory, consider what Strawson calls *P-predicates*. One of the principal thrusts of Strawson's argument is that P-predicates such as *is happy, is in love, is courageous*, and the like, are *not* reducible to M-predicates such as *is five feet tall* and *is dark skinned* since we need P-predicates as a basic category to make sense of our (social) world. However, Strawson does not deny that *the condition of application* of P-predicates is always routed via the availability of M-predicates; that is, we apply a P-predicate to an object only after identifying an object with an M-predicate even if implicitly: we apply *is smiling* to an object identified with *is a male*. This suggests a hierarchy of conditions without suggesting a *reduction* of one category to another. On this point, see also Hornsby (1998, p. 29).

But it seems counter-intuitive to think of happiness as a *property* of the individual so predicated; in fact an assembly of such 'properties', according to Strawson, does not individuate anything at all; that is why we need M-predicates to identify objects. To *what* then does the concept of happiness apply? Strawson does not say how this problem—that some concepts are at once basic and non-genuine—is to be addressed within the thick programme of descriptive metaphysics. Given the framework of ascriptions suggested above, it can now be said that some P-predicates articulate concepts which are functional only in their ascriptive role. In that way then the distinction between description and ascription is finer than the thick category of descriptive concepts Strawson is concerned with. Also, I am not suggesting that a taxonomy of ascriptive concepts will match Strawson's list of concepts articulated via P-predicates: *is schizophrenic*, for example, is certainly a P-predicate, but it is unclear as suggested above that it is not used with the intention to pick out a genuine property of something; *is divine*, on the other hand, is certainly an ascriptive concept without necessarily being a P-predicate: a piece of music or a glass of wine can be divine.

---

[3]For whatever it is worth, I guess Strawson might have used the term *description* in *descriptive metaphysics* following the influence of Wittgenstein's distinction between explanation and description, but he *meant* 'ascription' in the sense under discussion here.

Much of the above is likely to sound fairly obvious due to the paradigmatic nature of the examples chosen: beauty is certainly an ascriptive concept, spin descriptive. Having so anchored the distinction, it is interesting to ask whether there are concepts which are usually taken to be descriptive although they ought to be viewed (largely) as ascriptive concepts answering to our needs. For convenience, let us call them *hybrid concepts*. Where and how do we look for such concepts? The concept of God, briefly mentioned above, might begin to provide an answer. Even though the concept is commonly used to designate a supreme being, I suggested that the *history* of the use of the concept, which includes the turbulent history of the so-called ontological arguments, suggests that we are not describing anything at all; yet we are using the concept simply to display our (spiritual) needs, and a reverential attitude to the fact of creation.

However, the contours of the concept has a limited use for our purposes here. First, as any atheist will testify, it is in fact possible to do *without* the concept. This is because as parts of the fact of creation come within genuine understanding, the need for the concept diminishes. Second, even if we entertain spiritual needs, they need not be answered with the help of this concept: several religious systems do not have the concept. Third, it is false that those who do use the concept use it with identical needs and attitudes. Hence, the use of the concept is not sharp and stable enough to lay the basis for a general metaphysics of ascriptions focused on a massive 'ahistorical core'.

The concept of schizophrenia or the predicate *is schizophrenic* seems better suited for the job, although it has a rather restricted application. Thus, although the universal and regular availability of this concept may be in doubt, its stability in the cultures in which it is in fact (perhaps irregularly) available suggests a possible domain where we might locate a clustering of hybrid concepts. Note that the concept will be in use as long as perception of split-personalities is commonly available. Notice at once that even *this* concept is hybrid. It seems that here is a concept, split-personality, whose domain of application cannot be indicated without using other hybrid concepts; we must mention *persons*, that is, creatures *with a mind*, to indicate where the concept of schizophrenia applies. In this the predicate *is schizophrenic* closely resembles the predicate *is beautiful* since the domain of application of the latter concept requires mentioning *artistic* objects. However, *is beautiful* is not a hybrid concept; it is an ascriptive concept *par excellence*. Tentatively then, hybrid concepts seem to arise when the domain of application involves creatures with a mind since the concept of mind itself is hybrid.

Elsewhere (Mukherji 2000b, 2006; *this volume* Chaps. 8 and 9), I have argued that some of the central concepts of philosophy of mind such as knowledge and belief appear to be ascriptive concepts. Very briefly, it does not seem plausible to think of knowledge and belief to be *properties* of states of mind. But that ought not prevent us from *ascribing* knowledge and belief to cognizers. If these suggestions are valid, then the concepts of knowledge and belief are best viewed as hybrid concepts since they are typically *taken to be* descriptive concepts. Following this lead and combining it with the features of the uses of *is schizophrenic*, it follows that mental concepts, across the board, are likely to fall in the class of hybrid concepts.

With this, let us return again to some of the other features of the predicate *is schizophrenic*. I noted briefly that parts of the hybrid concept may in fact pick out some genuine properties of the objects under study. Professional psychologists hold the view that schizophrenia is an observed mental disorder with identifiable physical marks. In fact, professionals often *deny* that schizophrenia involves split personalities. Instead, they suggest that some forms of schizophrenia involve an imbalance of the brain chemicals or neurotransmitters: dopamine, glutamate and serotonin; obviously, such uses of the noun *schizophrenia* is no part of its common use.

Much more tentatively, similar remarks apply to the concepts of knowledge and belief. So perhaps the future history of these concepts justify why they have been hybrid ones, rather than fully ascriptive ones. But then large parts of these concepts continue to be prominently ascriptive, and these parts perhaps cover the most stable parts of their common use. In other words, the parts which do not lend themselves to scientific inquiry are exactly those which satisfy our needs and display our attitudes. If these observations about selective uses of 'mentalistic' predicates is credible, we might begin to understand where the contemporary divide between folk psychology and cognitive science is coming from.

7.5   Consciousness Revisited

We are now in a position to ask whether the concept of consciousness is a hybrid one. Several intuitions seem to cluster around this concept to suggest that it is indeed a hybrid concept. First, if anything, it is a mental concept designed to be used for creatures with a mind. It is no wonder thus that René Descartes took so much trouble to argue that animals do not have consciousness because he thought that they do not have a mind. Second, perhaps even more than the concept of God, the concept of consciousness is universally and ubiquitously in use because, as Chalmers put it, something is going on all the time, especially when we are awake. Third, it is typically taken to be a descriptive concept such that a lot of science goes into harnessing it; it is supposed to be the hardest problem faced by a putative science of the mind.

Following the lead from the concept of schizophrenia as our paradigmatic example of a hybrid concept, we might expect then that parts of the concept may in fact yield to science while the more stable and regulatively significant parts refuse to do so. To these latter parts, then, the notion of ascription will truly apply. In other words, the point of interest is to see whether the scientifically elusive parts in fact converge on what we take to be the real significance of the concept. To examine the possibility just suggested, let me dwell briefly on the idea that the problem of consciousness is the hardest problem for a science of the mind.

To my knowledge, it used to be said, at the early stages of cognitive science, that while consciousness is the hardest problem, meaning is the *second* hardest problem (Pylyshyn 1984, Chap. 1). It is interesting to note that not so long ago before that, these two were taken to be, roughly, the same problem. Recall, for example, that Rudolf Carnap (1932) thought of *erlebs*, or units of experiences in a time-slice, to furnish the basic units of meaning of, say, colour-words; the rest, for Carnap, was set theory. The actual mechanism suggested by Carnap fell into disrepute after Nelson Goodman's devastating criticism (Goodman 1951).

But the general idea of experiences supplying the 'basis' of meaning continued through the writings of Quine (1960) and Follesdal (1975), among others; for example, Quine's theory of language was based on the idea of stimulus-meaning. In the arena of Western analytic philosophy, to my knowledge, the issue finally came to a close with Donald Davidson's critique of the third dogma of empiricism (Davidson 1973). I am not so sure that the issue has been closed in other arenas such as discussions on Indian Theories of Meaning (*Nyaya* theory, in particular), and some versions of connectionist models of semantics currently in vogue (see papers by Davies, Ramsey, and Gärdenfors in Clark and Ezquerro, 1996). I have no intention of entering this debate in this chapter.

The only reason why I made a brief survey of the literature is to highlight the point that the idea that the problem of meaning is the *second* hardest problem is not immediately obvious. *Prima facie*, it makes a great deal of sense to say that the meaning of *cow* is essentially linked to experiences of cows; that's where meaning must ultimately be coming from. Then why is the problem of meaning taken to be the second hardest problem ranking *below* the problem of consciousness? The answer, in my opinion, maybe found in thinking of the concept of meaning as a hybrid concept. The notion of meaning that is entertained in the Carnap–Follesdal–*Nyaya* axis is essentially an ascriptive concept which in turn is linked to the ascriptive notion of phenomenal consciousness. I do not think anything decisive has happened in the meantime for us to locate *this* concept of meaning within science, notwithstanding tall claims from connectionist circles. Even if we bravely ignore Davidson's philosophical objections to the very idea of phenomenal meaning, the problem stays where it is. In a moment we will see why.

Yet true to the *hybrid* nature of the concept of meaning, some aspects of the original thick concept have indeed been brought under control. As quick examples, one may cite the work of Noam Chomsky and his followers in supplying a theoretical foundation to the notion of referential dependence (Chomsky 1995; Mukherji 2010), and the work of Davidson and his followers on truth-conditional semantics (Davidson 1967; Larson and Segal 1996). Personally, I have great difficulty in thinking of Davidson's work as contributing to the scientific study of meaning (Mukherji 2010), but the issue is irrelevant at this point. What is of interest is that neither of these approaches have anything to do with what we have called *phenomenal meaning* and, despite claims to the contrary, each of these approaches concern fairly remote, technical aspects of the problem of meaning far removed from our ordinary ascriptions of the concept of meaning. Our ordinary meaning of *cow* does concern phenomenal awareness of cows. In that the scenario parallels the history around the concept of schizophrenia. Thus, as far as the notion of *phenomenal meaning* is concerned, the problem of meaning is exactly as hard as the problem of consciousness.

It is easy to see why. The basic trouble is that the very nature of *phenomenal* consciousness, as distinguished from, say, informational consciousness and access consciousness (Dennett 1991a, b), is probably not open to the vocabulary of science, as we saw. To recapitulate, science necessarily invokes a third-person vocabulary in the sense that its results are open to public scrutiny, while phenomenal consciousness is essentially first person. Thus, Horwich (1998) has argued that truth-conditional semantics employs a third-person account. In contrast, following Nagel (1974), most researchers on phenomenal consciousness agree that a third-person account is at least elusive, thus opening an explanatory gap.

Phenomenal consciousness, along with phenomenal meaning, thus poses an apparently unsolvable problem. In this the problem is not unlike the problem of God. For something to be an account at all, we want the account to be empirically significant; yet, by definition, an account of God cannot be empirically significant since God cannot both be empirically significant, and be the cause for such empirical significance. In other words, we recognize something to be empirically significant because God caused it to be so; that's the concept of God we want. Our needs then, while raising the expectation for some account, prevent us from doing so.

As noted, the concept of God may in fact be given up precisely for the preceding impasse. Not so with the concept of phenomenal consciousness. Following Strawson, even if we apply P-predicates such as *is smiling*, *is in love*, *is in pain*, *is thinking hard*, and the like, to entities by first identifying them via M-predicates, we do not apply these P-predicates to all entities we identify via M-predicates; for example, we do not apply these P-predicates to trees and computers which we identify with M-predicates. Much of the groundwork for the functioning of language and society requires that we are able to form a conception of persons, and person-like creatures, after having formed a similar conception of ourselves first. The concept of consciousness plays a singular role in anchoring this conception of ourselves. No wonder, then, that the concept shows up the moment we wish to *extend* the concept of a person to fetuses, flora and fauna. Almost instinctively, we ask: is it conscious?

In a quick generalization, therefore, it follows that we need the concept to form some conception of an *ethical order* consisting of fellow beings just as we need the concept of beauty to form a conception of an aesthetic order. Those needs, it should now be obvious, are essentially normative, with no demand for descriptive truth; hence, there is no demand for a theory. Just as we may sometimes dispense with the concept of God, it is quite possible that we ought to be able to dispense with the concept of consciousness as well; every ascriptive concept must provide room for that possibility, provided that we are able to retain the notion of an ethical order *otherwise*. At the moment we have no idea of how to do so; so we retain the concept of consciousness.

References

Bennett, M.R., and P.M.S. Hacker. 2003. *Philosophical foundations of neuroscience*. London: Wiley-Blackwell.

Block, N. 2007. Consciousness, accessibility, and the mesh between psychology and neuroscience. *Behavioral and Brain Sciences* 30, 481–548.

Block, N. 2009. Comparing the major theories of consciousness. In *The Cognitive Neurosciences IV*, ed. M. Gazzaniga. Cambridge: MIT Press.

Carnap, R. 1932/1967. *The Logical Structure of the World*. Translated into English by R. George. Chicago: University of Chicago, 1967. Original German, 1932.

Chalmers, D. 1996. *The conscious mind. in search of a fundamental theory*. New York: Oxford University Press.

Chomsky, N. 1995. *The Minimalist Programme*. Cambridge: MIT Press.

Churchland, P., and T. Sejnowski. 1992. *The Computational Brain*. Cambridge: MIT Press.

Clark, A., and J. Ezquerro (Eds.). 1996. *Philosophy and Cognitive Science: Categories, Consciousness and Reasoning*. The Netherlands: Kluwer.

Cottingham, J., R. Stoothoff, D. Murdoch, and A. Kenny (Trans.). 1991. *The Philosophical Writings of Descartes, Volume III: The Correspondence*. Cambridge: Cambridge University Press.

Davidson, D. 1967. Truth and meaning. In *Inquiries in Truth and Interpretation*. London: Blackwell, 1984.

Davidson, D. 1973. On the very idea of a conceptual scheme. In *Inquiries in Truth and Interpretation*. London: Blackwell, 1984.

Davidson, D. 1975. Thought and talk. In S. D. Guttenplan (Ed.) *Mind and Language*. Oxford: Clarendon Press.

Dennett, D. 1991a. *Consciousness Explained*. Boston: Little Brown.

Dennett, D. 1991b. Mother nature versus the walking encyclopedia. In *Philosophy and Connectionist Theory*, S. Stitch and D. Rumelhart (Eds.). Earlbaum.

Flanagan, O., and D.T. Dryden. 1998. Consciousness and the mind: contributions from philosophy, neuroscience, and psychology. In *Methods, Models, and Conceptual Issues: An Invitation to Cognitive Science*, D. Scarborough, and S. Sternberg (Eds.). Cambridge: MIT Press.

Fodor, J. 1980. Methodological solipsism considered as a research strategy in cognitive science. *Behavioral and Brain Sciences* 3, 63–73.

Fodor, J. 1983. *Modularity of Mind: An Essay on Faculty Psychology*. Cambridge, MA: MIT Press.

Follesdal, D. 1975. Meaning and experience. In *Mind and Language*, S. Guttenplan (Ed.). Oxford: Oxford University Press.

Goodman, N. 1951. The problem of imperfect community. In *The Structure of Appearances*. Cambridge: Harvard University Press.

Goodman, N. 1978. *Ways of Worldmaking*. New York: Harvester Press.

Hornsby, J. 1998. *Simple Mindedness. In Defense of Naive Naturalism in the Philosophy of Mind*. Cambridge: Harvard University Press.

Horwich, P. 1998. *Norms of language. In Meaning*. Cambridge: Cambridge University Press.

Larson, R., and G. Segal. 1996. *Knowledge of Meaning*. Cambridge: MIT Press.

McGinn, C. 1989. Can we solve the mind–body problem? *Mind*, ICVIII, 391, July.

Mukherji, N. 1990. Churchland and the talking brain. *Journal of Indian Council of Philosophical Research*, 7(3, May–August).

Mukherji, N. 2000a. *Cartesian Mind: Reflections on Language and Music*. Shimla: Indian Institute of Advanced Study.

Mukherji, N. 2000b. Traditions and concept of knowledge. In *Science and Tradition*, B. Patnaik, A. Raina, and M. Chadha (Ed.). Indian Institute of Advanced Study: Shimla.

Mukherji, N. 2006. Beliefs and believers. *Journal of Philosophy*, Calcutta University, December.

Mukherji, N. 2010. *The Primacy of Grammar*. Cambridge: MIT Press.

Nagel, T. 1974. What is it like to be a bat? *The Philosophical Review* LXXXIII(4), 435–50.

Penrose, R. 1994. *Shadows of the Mind*. Oxford: Oxford University Press.

Putnam, H. 1981. *Reason, Truth and History*. Cambridge: Cambridge University Press.

Pylyshyn, Z. 1984. *Computation and Cognition*. Cambridge: MIT Press.

Quine, W. 1960. *Word and Object*. Cambridge: MIT Press.

Ryle, G. 1949. *The Concept of Mind*. Chicago: University of Chicago Press.

Searle, J. 1992. *The Rediscovery of the Mind*. Cambridge: MIT Press.

Singer, P. 1976. *Animal Liberation*. London: Jonathan Cape.

Stich, S.P. 1983. *From Folk Psychology to Cognitive Science: The Case Against Belief*. Cambridge, MA: Bradford Books/MIT Press.

Strawson, P. 1957. *Individuals*. London: Methuen.

Strawson, P. 1992. *Analysis and Metaphysics*. London: Oxford University Press.

Tononi, G., and C. Koch. 2015. Consciousness: here, there and everywhere? In *Philosophical Transactions of The Royal Society*, Published 30 March 2015.

Wang, H. 1987. *Reflections on Kurt GÖdel*. Cambridge: MIT Press.

Chapter 8
Ascription of Knowledge

> *And the smallest class is the one which naturally possesses that form of knowledge which alone of all others deserves the title of wisdom.*
>
> Plato

There is some justification to an increasingly popular view, especially among scholars in the humanities and the social sciences, that some aspects of modern science are problematic, particularly in the Indian, or, more accurately, in the developing context. One perceived problem is that, instead of appearing as a messenger of enlightenment, modern science often assumes the form of hegemony.

In the Indian context, the complexity and power of modern science is somewhat more acutely felt since most knowledge enshrined in modern science is imported from the West, and is thus alien to the traditional ways of thinking. There is a sense of interference and imposition, and a feeling of apparent lack of control over indigenous lives. Needless to say, many realms of thought and action are needed to understand and confront this problem; for example, we need to address confusions about the nature of science, routine failure to make a distinction between basic science and technology, and the like. Yet, grim issues do arise to the point of threatening the survival of the species. Here my goal is more modest: to understand the relation between modern science and Indian culture.

8.1 Traditions and Categories

In this chapter, instead of engaging with such issues in the sociology of science, I will attempt to isolate some theoretical features of the problem to understand it in terms of my own professional training. Along the way, I will throw up some hints as to why certain popular conceptions of the issue fail to bring out deeper conceptual concerns. My general aim is to establish some links between the social

---

This is a revised version of a paper published as Mukherji (2000).

© Springer Nature Singapore Pte Ltd 2017
N. Mukherji, *Reflections on Human Inquiry*,
DOI 10.1007/978-981-10-5364-1_8

aspects of the study of knowledge-systems and Western analytical epistemology. These two branches of inquiry are seldom placed side by side; in fact, the received image is one of studied indifference towards one another. The qualification 'analytic' is significant because there is growing interaction between social theorists and philosophers of other hues such as Buddhists, Jainas, and continental philosophers.

Analytic philosophers do not have a very charitable view of such interactions. An analytic philosopher is primarily concerned with conceptual questions regarding the necessary and the sufficient conditions of knowledge: what are the defining features of the concept of knowledge such that it is both rational and commonly held by people? The concern is self-consciously universalistic and, thus, it is apparently devoid of historical content. Hence, the inquiry is hardly concerned with the context and growth of specific knowledge systems as they change our lives.

A sociologist of knowledge, in contrast, is directly concerned with the domain just mentioned: what social, political, and cultural factors allow or disallow a certain knowledge-system to emerge in a certain context? What are the boundaries and the institutional features of this system? The inquiry here need not await clarification of the very concept of knowledge and, thus, of the concept of knowledge-systems. Knowledge-systems are there for all to see: folk-theories, theological texts, modern science. They arise and grow apparently in specific historical contexts, and guide our lives. The task of the sociologist is to detail the nature of this guidance. Hence it is no wonder that sociologists and analytic philosophers seldom look at each other.

A possible myth underlies these disciplinary distinctions: the myth that 'pure' conceptual inquiry cannot itself be placed under sociological examination. One may legitimately seek the universal features of a concept, but the search itself must have historical features which, perhaps, are linked to specific traditions; so, the search becomes an element in the class of knowledge-systems the sociologist wants to examine. To that extent, the concept so examined becomes tinged with traditional colour. For example, it could be that some conception of (spiritual) liberation is a part of every human culture; yet, the Indian concepts of *mokṣa* and *nirvāṇa* are certainly rooted in that culture.

I am not suggesting that this ought to be the case, that every universal search is somehow self-defeating. It is quite possible that there are universal features in methods of inquiry such that when they are directed at putative universal phenomena, both the inquiry and its object highlight each other's universal features. Mathematics and related disciplines could be cases in point.[1] Even though a certain form of geometrical knowledge was developed in ancient Greece, there seems to be nothing particularly 'cultural' in the knowledge that the three internal angles of a Euclidian triangle add up to 180°s. In turn, the inquiry into what form of thinking generates such abstract truths—the celebrated Hilbert's programme—was also non-cultural in that sense even if it originated in Europe. Things might just so

---

[1] I am compelled to cite some obvious cases at this point because I have been told by fairly enlightened professional thinkers that, say, modern physics is just a 'vocabulary' of the Europeans.

converge with universal content in the favourable cases. Such, it has been claimed, is the common ground for much of analytic philosophy *and* modern science. The perspective was examined in great depth, for the first time, by Immanuel Kant in his *Critique of Pure Reason* (Kant 1787).

Still it is hard to see how an *evaluation* of whether a universal discourse has been reached is itself free of traditional outlooks, however widely shared; the Hilbert programme was a European programme after all, not to mention modern physics. So, although it is conceivable that the complex problem of how we are able to theories about contexts-in-general, while occupying only one of them, gets *naturally* solved from time to time, it is far from clear that we could make a genuine *claim* to that effect. In other words, it is one thing to simply reach a universal concept, or even a form of inquiry, without noticing it as it were; it is quite another to declare so. That additional step seems to require marshaling of historical arguments to show why the inquiry in this particular case assumes an unhistorical character.

These reflections cast doubt on the so-called 'enlightenment' view that modern science, or any knowledge-system for that matter, is a truly universal achievement; to repeat, the doubt is not necessarily directed at the content of modern science, but at the form of knowledge-system itself. The knowledge-system of modern science —its proofs and evidence—may well be universal, but then explicit argumentation is needed to show that the foundational elements of modern science transcend traditional boundaries. If, in contrast, explicit argumentation is indeed marshaled to the effect that some of these crucial elements are themselves embedded in specific traditions, then the universality, and therefore the universal goodness, of modern science will be seriously challenged. Why just modern science? Even basic aspects of rational thinking itself could be challenged; for example, sophisticated scholars argue that deductive thinking—arguing from premises to conclusions in terms of a fixed set of rules—is an unfounded principle of the occidental culture (Raju 2007). It is another matter whether such critics are themselves able to avoid the forms of inquiry they are challenging.

8.2 Form of Explanation

The issue, just raised, is particularly significant for the Indian tradition that otherwise displays immense intellectual cultures in areas other than modern science. If a tradition failed to display sophisticated intellectual cultures similar to that of modern science, then it would have been possible to uphold something like a 'stage-theory' of traditions in which traditions go through stages of development to finally reach the pinnacles of, say, modern science. We could say that the European tradition has already reached it, and other traditions—when suitably nourished perhaps with inputs from the European tradition—will follow suit.

This view, held by the perpetrators and victims of colonialism alike, simply does not work for the Indian tradition. The fact that extremely complex and abstract theories of language and tonal music—not to mention elaborate philosophical systems—were formulated in this tradition without leading to modern science much before such theories were even contemplated in the West, is a puzzle that no 'stage-theory' can explain. Moreover, in the West, sophisticated theorizing in *these* domains took place *after* the emergence of modern science. In the case of language-theory, the first Western theory that could probably match the intellectual sophistication of Pāṇini (350 BC) was developed only in the thirteenth–fourteenth centuries in the form of *Modistic* grammars (Hinzen and Sheehan 2013); in fact, it is possible to argue that Paninian grammar was finally matched only in the 1950s in the form of Chomskyan linguistics. Notice also that the wide separation in history and geography between Pāṇini and Chomsky does not prevent us from viewing either Pāṇinian theory or Chomskyan theory as universal theories of human language (Mukherji 2002; *this volume*, Chap. 6). So the picture of linear growth of all traditions towards a universal destination is faulty at more than one place. What, then, explains the *absence* of modern science in the Indian tradition? A clear answer to this question is crucial for deciding the stand we should adopt towards modern science.

The preceding way of setting up the problem already rules out certain facile answers and attitudes. For example, it has been suggested that the Indian tradition did not develop mathematical theories although it did develop complex computational techniques, or that it did not encourage disinterested empirical enquiry, or both. Even if these suggestions are empirically valid (which is seriously doubtful), it is obvious that they simply repeat the statement that modern science did not develop in India; they don't *explain* the statement. Why didn't mathematical theory develop in India even though the ancient Indian tradition did produce supreme formal thinkers such as Pāṇini and a long line of his followers in the grammarian tradition, and a range of logicians in a variety of philosophical schools?

Also, it will be too facile to blame certain cultures within the tradition, such as the *Brāhminical* culture, for *resisting* the emergence of science. This is an European model of looking at history: the medieval scholastic tradition lost their resistance to modern science in Europe, while the traditionalists (allegedly) won in India. Even without recommending the tenets of *Brāhminical* culture, whatever they are, it is simply false that events parallel to European history ever took place. *Brāhminical* culture was not the only intellectual culture in India, and even within the *Brāhminical* culture there were any number of schools of thought that challenged each other's foundational positions over centuries (Sen 2005). How come the preparations for modern science singularly escaped this vigorous and enlightened debate that covered almost every aspect of human society and knowledge? Similar remarks apply at least to the Chinese and Islamic traditions.

The fact of the matter is that there never was any resistance to modern science simply because modern science didn't even appear on the scene for someone to resist it. It seems then that modern science is somehow *intrinsically* alien to the Indian tradition much like the separation of branches in the tree of evolution. There must be some radical difference between the traditions such that the conditions for *specific* knowledge-systems are not even available across them. Unless we clearly understand these differences at the correct level of description, we might be tempted to embrace sweeping ideological positions that reject the value of modern science itself just because it is historically alien to non-Western thinking. In the opposite direction, an uncritical acceptance of modern science could encourage 'racist' views of other cultures. Setting the political issues aside for the moment, we cannot even make sense of these views unless we know where exactly the traditions fail to fit, before we begin to attach value to this lack of fit.

I am compelled to talk of 'radical' differences between traditions in view of the empirical nature of the problem, namely, the striking absence of entire knowledge-systems in specific cultures. How radical? A fair amount of loose thought hovers around this question, in my opinion. In looking for categories and concepts that might be used to fundamentally distinguish between differing traditions, we *cannot* dig up categories which attach to the very genetic make-up of people. Notice that *any* set of such categories will lead to the obviously false conclusion that modern science is unlearnable by those people. Since there are no such people, there aren't such categories.

The argument here is analogous to the refutation of some recent claims about the speakers of pirahã. Some linguists have claimed (Everett 2005) that the pirahã language does not have grammatical recursion because speakers of pirahã do not comment on anything except 'immediate experience;' their world-view apparently consists of 'here and now.' The problem is that, notwithstanding the attractive world-view, the claim implies that speakers of pirahã cannot learn a language which has grammatical recursion, a claim that is patently false. Apart from drawing attention to a related analogy, I mentioned this case here because it is popular these days to trace 'ontological' issues to 'linguistic' ones, and vice versa. In searching for differing categories, one cannot go as far as to list *ontological* categories— world-views—to show differences between traditions since ontological categories are *too* deep-rooted.

In any case, as Davidson (1973) has shown, the idea that people differ radically in their ontological categories (or that they don't) is self-defeating. The idea can be posed only if the categories concerned are translatable in the allegedly differing schemes; but, if they are translatable, they cannot differ radically. And there is no third scheme in which it can be shown that the two earlier schemes are untranslatable; for the very showing of 'untranslatability' will identify a scheme in which both categories can be held. Similar remarks apply to those sociologists who trace the differences between 'cultures' to fundamental differences in global concepts such as rationality. But sociology of knowledge is a fast-track bestselling discipline that has no patience for subtle philosophical argumentation.

Nevertheless, the concept that marks a tradition must be foundational enough to survive through millennia of natural pressures, assimilations, and change. An optimal solution to these somewhat opposing requirements will display categories or concepts that are parametric in nature: the concept is available across traditions, but the specific form (including the *null* form) in which it is realized depends on the specificities of the tradition at issue. In other words, these methodological points demand that there ought to be some deep-rooted concepts in the Western tradition which are specific to that tradition, and may be absent in other traditions, without affecting the genetic make-up of people. The presence or absence of modern science could perhaps be traced to such concepts. This is where, in my view, sociology of knowledge can gainfully meet analytic philosophy. Having thus outlined the contours of the concepts we are seeking, where are we likely to find them? My professional instinct suggests that one ought to look closely at the philosophical cultures within traditions. This is because philosophers are most likely to isolate and examine individual concepts of a fundamental sort to furnish a general account of human thought to highlight the universal validity of their own traditions.

8.3   Concept of Knowledge

In what follows I argue that one classical and influential version of the concept of knowledge—*epistēmē*—vigorously examined in the Western tradition at least since Plato, fits these requirements just outlined. There are other related concepts that satisfy these requirements. Elsewhere, I have argued that the concept of belief has, unsurprisingly, similar effects (Mukherji 2006; *this volume*, Chap. 9). However, the contours of the two concepts also differ quite sharply at many points; hence, it is advisable to keep them separate for the moment.

We will see that the concept of knowledge is best understood primarily as a socio-normative concept without any psychological underpinnings. I am not suggesting that this is how the concept is in fact understood in the Western tradition. It is not even clear that Plato would have allowed a psychological/social distinction around this concept. Although the normative nature of the analytic concept of knowledge is sometimes recognized in the contemporary literature, there is a fair amount of confusion as to how to accommodate this fact with the allegedly psychological nature of the concept. Most authors believe that the concept is somehow a mixture of the two, without telling us how these apparently disparate elements can be mixed at all, and leave matters at that. I do not know of any significant literature that argues for the primarily non-psychological nature of the concept.

I will argue that *unless* we understand the concept of knowledge normatively, it turns out to be nearly vacuous. I will concentrate on this crucial point for much of the rest of the chapter. Once the normative nature of the concept is clear, it might be possible to relate it to specific traditions, although this step will require additional empirical evidence.[2] The final step will require that the concept, so placed in a tradition, is shown to be crucially linked to modern science. Except for making some intuitive remarks, I will not go into the last two steps of the argument in this chapter. I hope the motivation and thus the plausibility of the entire project gets clear once the normative character of such a foundational concept is fully displayed.

The study of this concept is directly relevant for the project at hand, since recent work suggests that the concept of knowledge, as discussed in the Western tradition,

---

[2]This caution is needed in view of important work by Hilary Putnam (1994) who, following Kant, has argued that normativity need not be opposed to universality.

never played any significant role in the Indian philosophical tradition (Mohanty 1992). The Indian tradition abounds with discussion of the concept of *jñāna* whose nearest English equivalent would be *cognition, not* knowledge; what *jñāna* means is something like the moment or occurrence of enlightenment. This is a scholarly point whose establishment is beyond the scope of this chapter; hence, I will simply take it for granted. Nevertheless, whatever be the details of the concept of *jñāna*, it is clear that it is meant to pertain to states of individual cognizers, although it is doubtful, in the absence of a clear mind/body divide in the Indian tradition, whether these states should be construed as *mental* states in the Western sense.[3]

Now, if the Western concept of knowledge also applies to states of individuals and the concept is absent in the Indian tradition according to scholars, it will follow that the two traditions postulate radically different views of individual (mental) states; the members of the two cultures differ in their basic mental make-up. This consequence appears to violate the methodological restriction sketched above. I am able to avoid this unsavory consequence by arguing that the concept of *knowledge* has nothing to do with states of individuals. Could it be that the absence of concept of knowledge in the Indian tradition is linked to the absence of modern science in some way? Notice that the concept can be linked, if at all, with modern science only necessarily, not sufficiently. Just the upholding of a concept is not enough to construct large-scale knowledge-systems; for that to happen, other personal and institutional resources are also needed.

To see how the concept functions, let us go back to Plato's original discussion. Plato invites us to imagine some people in a cave. I will use a bit of my own imagination to highlight Plato's central point. There is a wall, in the otherwise dark cave, which is partly and dimly illuminated by light from an opening such that shadows of fleeting external objects are projected on it. The people face the wall with their back to the opening. Shadows fall on the wall when objects momentarily pass. The conditions are such that: (a) people watch only the shadows of these passing objects on the wall; and (b) their own shadows get mixed up with those of the objects. In that sense, the objective understanding of the world is mixed up with subjective content (see *this volume*, Chap. 2).

Plato construes this sort of understanding of the external world as a set of *beliefs* where there is no clear distinction between how objects really are, and how they appear to people. People get *knowledge* when they leave the cave, step outside, and look at the external world in direct sunlight. Let me note in passing that, since the allegory appeals to the analogy of vision, it naturally promotes a psychological view of human knowledge. However, the visual analogy is just a handy tool, it is not intrinsic to the issue.

Much has been made of this and related allegories—for example, the allegory of line—in the literature. There are two crucial features of knowledge, according to

---

[3]The concept that carries more of the sense of *knowledge* is possibly *pramā*, the state of true cognition (Mohanty 2000). Even though the concept of truth (or validity), *prāmāṇya*, is included in the definition of *pramā*, it is unclear if it applies to propositions, as in the Western tradition, or to valid states of cognition directly, say, in terms of sameness of representation.

Plato. First, knowledge is a *species* of belief, that is, every item of knowledge is also an item of belief, but not vice versa: some beliefs, in fact a large majority of them, fail to be upgraded to knowledge. Secondly, beliefs are unclear, local, momentary, and rather subjective in character; knowledge is clear, global, stable, and objective. There is a very large agreement in the literature that these features may be captured in the following *criteria of knowledge*.

For an agent S and a proposition *p*, S knows that *p* just in case:

1. S believes that *p*,
2. *p* is true,
3. S is justified in believing that *p*.

In other words, knowledge is viewed as *justified true belief* (JTB). Supposing this scheme to have captured Plato's idea of knowledge, how is this scheme to be understood? What is it an account of?

It is standardly thought that the criteria of knowledge laid out above gives an account of someone's coming to have a piece of knowledge. In other words, it is an account of what it is for S to know something: S knows something just in case S satisfies the criteria set above. Although it does not yet logically follow, it is thought that the criteria describes S's mental state when S comes to have a piece of knowledge. This view is perhaps enhanced by the thought that both having a belief, and having some evidence/justification for that belief, are apparently mental acts and states themselves. This takes care of conditions (1) and (3). Condition (2), on the other hand, says something about how the world is, namely, it is such that a certain state of affairs, as enshrined in the proposition *p*, obtains. S attains knowledge when his mental state not only matches the world, S is also aware that it does. Thus the criterion is understood in purely psychological terms. All of this is at least compatible with Plato's allegory.

The criteria, as stated above, immediately rules out the application of the concept of knowledge in a large variety of cases. The criteria are restricted solely to conscious propositional knowledge since the question of having some justification does not arise unless the agent is conscious. As such, the criteria does not apply to any of the following cases:

a. Unconscious non-propositional knowledge,
b. Unconscious propositional knowledge,
c. Conscious non-propositional knowledge,

and any variety within each of (a), (b) and (c). Thus, to mention just a few cases: it does not cover cases of *nirvikalpaka* (non-conceptual) cognition argued for in much of Indian epistemology; tacit knowledge of language; repressed knowledge studied in psychoanalysis; knowing how, knowing who, knowing when, knowing what, and the like. And, of course, the JTB conception of knowledge cannot apply to nonhuman animals since animals cannot be said to have a propositional attitude (Davidson 1975); Pranab Sen (2007, pp. 27–28) finds this to be 'extremely presumptuous.' The exclusion of *nirvikalpaka* cognition is interesting for the issue in hand. For those schools of Indian philosophy, such as *Nyāya* and kindred systems, *nirvikalpaka* cognition is the foundation on which all cognition is built up. So there is a direct lack of fit between the JTB conception of knowledge, and some major Indian traditions.

Apparently, then, the JTB conception applies, if at all, only to cases of propositional knowledge for which we have (conscious) evidence. Having thus isolated a domain of application, contemporary analytic philosophers have been exclusively concerned with the issue of whether the individual conditions mentioned in the criteria are (jointly) sufficient for the concept of knowledge; that is, whether the JTB conception can be viewed as a definition of knowledge. Following the immensely influential work of Edmund Gettier (1963), it turned out that many beliefs fail to qualify as knowledge even when they meet JTB-sufficiency. Hence, the post-Gettier attention on JTB is squarely focused on the issue of how to formulate some additional conditions on knowledge. It is taken for granted that the set of conditions (1)–(3) are individually necessary for the restricted domain of application.

Recent work by Sen (2007), in contrast, raises doubts as to whether the conditions are even individually necessary. Since Sen's work is not much known outside a handful of devoted scholars in India, it is worthwhile to review his complex and important reflections on the Gettier conditions. Sen begins by suggesting that the conditions (1) and (2) of JTB are not independent; one cannot be justified in believing that *p* without believing that *p* (Sen 2007, 29–30). Therefore, once the second condition is listed, the first ought to be deleted. However, given the basic Platonic condition that knowledge is a species of belief, we might wish to state the belief-condition explicitly, and hence retain (1). Suppose then we delink the justification-condition from the belief-condition somehow, as Sen proceeds to articulate. However, I will set this reconciliatory move aside since it is unclear what the *object* of justification in that case is, if it is not belief.

But can we do the reverse? Can we say that separate mentions of belief and justification are not needed because the conception of belief *includes* the conception of justification? Sen (2007, 31) does hint at the option without developing it. What it is then for someone to hold a belief in the first place, say, the belief that babies cry when hungry? Why should the belief be held at all? There ought to be *some* evidence, *some* justification for a belief to enter our heads. The point is: it is counter-intuitive to think of *holding* a belief that is *totally* devoid of any evidence/ justification. Surely, there is a clear distinction between our recognition of a grammatically meaningful proposition, and believing the proposition. I simply do not hold the belief that tigers, not elephants, have trunks; nor do I believe that I do not have a right hand. I do not do these things plainly because I have absolutely no warrant for doing so.

So to uphold the notion of entertaining a belief without any justification whatsoever is to get the psychology completely wrong. Of course, the justification offered may be inadequate, or downright wrong, or may be disputed. When that is pointed out to the concerned subject, she is likely to come up with another justification, including the 'mere' justification that she prefers to believe so. I once heard a candidate for a position in philosophy declare that God exists, whether we believe it or not. It seems then that the concept of justification is built into the notion of holding a belief that *p*, or of believing that *p*; in other words, the second condition is built into the first.

Consider next whether conditions (2) and (3) are individually necessary, that is, whether *both* the conditions of truth and justification are independently motivated. As noted, and as Sen points out immediately (2007, 33), there is a vast difference between the character of (2) and (3). While (3), the justification condition, may be said to be subject-dependent—in fact, it is a part of the subject's belief as argued above—(2) is proposition-dependent. So, it is at least odd to suggest that it is a necessary condition for S's knowledge that a proposition of a *language* is true. Why should my knowledge be dependent on properties of linguistic items? I may add that, given the correspondence theory of truth (*this volume*, Chap. 2), it is also odd that *my* knowledge depends not just on *my having* knowledge, but how the world is like independently of me. In other words, it is problematic that a description of *my* mental state includes a mention of how the world is like.

In line with these oddities, Sen points out another formidable problem in the idea that (2) and (3) are independent of each other. Given the preceding oddity about mixing subject-dependency with world-dependency, how can justification be independent of the truth-condition if the knowledge is to be *ascribed* to subjects? According to Sen (2007, 36), the problem is most acute for first-person ascriptions of knowledge:

For me, to decide whether or not the proposition I believe is true is the *same* as deciding whether or not I would be justified in believing it.[4]

Suppose we develop the general point as follows. Using the notions of justification and evidence as roughly equivalent for the purposes of this chapter, one could say that S is justified if S has some evidence. Now, it is obvious that to have some evidence for *p* is to know that *p* is true; how else could we have evidence that pertains to *p*? What does it mean to have evidence *for p* when *p* in fact is false?

Consider, for example, instances of knowledge such as:

- that grass is green;
- that babies cry when hungry;
- that most tigers have stripes;
- that I have a right hand.

What counts as evidence for, say, that babies cry when hungry? It is the evidence that babies cry when hungry. When the evidence is consciously entertained, then, given assumptions about enough linguistic knowledge, we cannot fail to know that the sentence *babies cry when hungry* is true. Essentially, the truth of *p* is built into the concept of evidence for *p*, insofar as the mental make-up of the subject is concerned. It seems then that at least one of the conditions (2) and (3) is redundant.

---

[4]I think Sen's observation is supported from a slightly different direction by considering why truth 'governs' beliefs in the first place, especially for first-personal doxastic deliberations (Shah 2003). I must mention that although Sen's paper mentioned here was published in 2007, he delivered it as the S. K. Sen Memorial Lecture in Delhi in 1997. Sen died in 1999.

It is plausible to think that (2) is redundant since (3), the justification clause, apparently carries a mentalistic dimension that is absent in (2) (giving rise to the oddities above), while the factual/evidential dimension of (2) is already captured in (3). As we saw, (3) is already included in (1), the belief clause.

To know that p, then, is simply to obey condition (1), namely, to believe that p. Recall that one crucial feature of Plato's theory was to capture the *asymmetry* between knowledge and belief. That feature collapses. To put this conclusion in another way: JTB seems to have no application at all, not even in its favored restricted domain.

8.4 The Knowledgeable

Should we just abandon the JTB conception of knowledge? What happens then to the allegory of the cave? Was Plato mistaken in the way he viewed human knowledge? Unfortunately, Sen does not enter into these questions. It seems to me that, despite pointing out the preceding difficulties with JTB, Sen is not prepared to give up the idea that JTB does capture some of the significant features of the *mentalistic* concepts of belief and knowledge. Throughout the complex discussion, Sen maintains that the JTB, as it is classically formulated, is in need of much amendment if it is to qualify as an adequate theory of the mental states of belief and knowledge of a subject.

In my opinion, Sen's arguments in fact show that, as a psychological theory, JTB is quite hopeless. It simply does not furnish any cogent account of what it is for someone to know something. Yet that is *all* that Sen's arguments show as they culminate in what he calls the 'devastating consequence' for first-person ascription. They still leave the possibility that Plato's theory and JTB be viewed in essentially non-psychological terms.

It is of some interest to note at this point that knowledge, in the JTB sense, is said to be an ascription. Although this term is used frequently in the literature, its unique connotation is seldom probed. Presumably, most authors simply take it for granted that to ascribe knowledge (or belief, for that matter) is to say that an agent *has* some knowledge—that is, that S knows that p—and JTB gives the condition of such knowing. What is missed here is the possibility that the JTB conception is viewed just in terms of conditions for ascription, not in terms of conditions for having knowledge. Since I have already discussed the distinction between ascription and description in detail earlier, in the context of ascription of consciousness (*this volume*, Chap. 7), I will assume the distinction in what follows.

The distinction between ascription of knowledge and having knowledge is pretty obvious. *Having knowledge* is a one-place predicate satisfied by a single argument, namely, the agent S. *Ascription* (or the verb *to ascribe*), on the other hand, is at least two-place: *someone* ascribes knowledge to *someone* (else). So ascription is an interpersonal concept, which may apply reflexively to oneself as well. Moreover, A may ascribe some knowledge to B without B's having that knowledge. In contrast,

B may have some knowledge without anyone ascribing it *to* B, including by B. This was Socrates' midwifery case, in which Socrates proceeded to dig up, as a 'midwife', some piece of knowledge entertained by the slave-boy; even the slave-boy did not know that he has had this knowledge until Socrates intervened. Given the clear distinctions between the two notions, it seems reasonable to attach JTB to one notion without attaching it to the other.

Let us, therefore, use JTB as a criteria not of someone having knowledge, but of ascribing knowledge to someone. Notice the moral tone already building up in the analysis. Plato's theory may now be thought of as laying down the condition for ascribing knowledge which, in effect, amounts to granting *authority* to the person. When the conditions are satisfied, the agent may be viewed as a *knowledgeable* person, someone who can be relied upon to spread wisdom. The interesting question, in the light of the preceding discussion, is: what is the domain of application of this concept of knowledge?

In most cases of ordinary cognitive life of humans, the question of granting authority or wisdom simply does not arise. Take for instance the items of knowledge listed above: that babies cry when hungry, that tigers have stripes, that I have a right hand, and the like.[5] These are items of *common* knowledge. Everyone *has* such items or, at least, they can be attained, pathology aside, by anyone in the natural course of living. They relate to our daily practices that are deeply embedded in the world of evidence. We attain these items by simply immersing ourselves, largely unconsciously, in everyday practices. Let us reserve the handy word *cognition* for this active area of human life. The activity *comes fully linked to evidence*; no further justification is needed, no truth-claim needs to be separately investigated. It is no wonder that the JTB conception of knowledge failed to apply to the complex labyrinth of our common cognitive life. This life still requires an account, a scientific one, that continues to elude us.

The JTB conception, on the other hand, is not an account of anything at all. It concerns our attitude towards things and processes not encountered so far—things that are uncommon, that are not routinely encountered, and thus fall outside the scope of the complex labyrinth just sketched. They cannot arise from the natural course of largely unreflective, daily practices; hence when they arise, we do not have any secured natural responses to them. Consider, for example, statements such as

- God exists.
- The Sun goes around the Earth.
- Animals and humans have a common origin.
- Human history is laced with violence.
- The Mind and the Body are separate substances.
- The Universe is filled with dark matter.

---

[5]It is obvious that this is just a suggestive list. The list will vary according to the varying conditions of cognizers. But it is not ruled out that humans, being humans, share vast areas of common knowledge. See the discussion on modularity in the next chapter (Chap. 9).

These are big, somewhat unwarranted and, therefore, surprising claims. For the purposes of this chapter, I will call them *texts*. Suppose someone comes up with a text: how do we respond to the person, to the text? What rational attitude should we adopt?

As the tentative and hopelessly incomplete list of texts shows, they are the natural ground for much of theology, metaphysics, and science. Since we have no natural responses to them, it is important to set up some normative criteria for evaluating them. We need to find out if the knowledge-claims enshrined in the texts are true. They cannot be *plainly* true since they do not arise naturally. So there must be some rather complex ways of determining their truth, referring perhaps to an intricate evidence-set, or some already-acknowledged text, or some rigorous application of mind, or all of these together.

Given the complexity of the truth-part of the claim, it cannot be immediately granted to the agent, even if the claim turns out to be true on independent grounds. The agent must satisfy the condition that he has access to these grounds; that he is justified in holding the belief. Only then the agent may be granted/ascribed knowledge of the text. Notice the feature of intellectual responsibility here. Once the knowledge is ascribed to the agent, the agent's word will spread the text henceforth, just as the truth-part of the current claim has already referred to such words of historically notable mouths.

Ascription of knowledge thus primarily concerns convincing others on matters of which we only have a dim view. The more sceptical the ascriber, the more complex the process of ascription. In the extreme case, where a minimal reference is made to other texts in the course of truth-evaluation, the ascription hinges on complex and remote evidential and logical resources. This requires going deeper into the nature of things to see if the text on the table holds good. Other texts are created along the way, and the process of evaluation gets even more complex. We demand complicated 'ontological arguments' for *God exists*, and an entire theory of evolution for *Animals and humans have a common origin*; yet we continue to withhold ascription.

In the other extreme of mutual agreement, where the understanding of the general nature of things is already deeply enshrined in some sacred text, and where both the agent and the ascriber are equally committed to the prevailing text, ascription takes a minimal form, perhaps no form at all. In a way, then, we can think of traditions in which no knowledge-*claims* of the textual sort are made any more. Statements from prevailing texts are simply repeated (or, perhaps, continuously clarified) in this tradition, since, in some sense, *all* knowledge has already been reached. These are, of course, idealized extremes to bring out the entire spread of human knowledge. In practice, knowledge-claims and ascriptions fall somewhere between mounting incredulity and immediate agreement.

The dilemma that ascriptions address is a curious one. On the one hand, ascriptions grant authority to an agent; on the other, just prior to the ascription, this proclaimed authority is relentlessly questioned. Authority is granted to those who can win the reflective confidence of others. As more agreement is (rationally) reached, the more the knowledge is shared, and a new text comes into being. The task, thus, is to reduce the effect of particular human agency continuously in an attempt to eliminate it altogether to reach what Thomas Nagel (1986) has called the *view from nowhere*. Recall that the removal of the subjective element was one of the central features of Plato's theory. Call this the search for *truth*, or *objectivity*, or whatever. By the very nature of the process, it can never be fully attained, but preparations can be made for its enlargement in the next stage. Traditions that do not explicitly entertain the ascriptive concept of knowledge thus apparently face the problem of depending uncritically on human agency. In that sense, they apparently miss out on truth and objectivity.

This problem, however, loses its force if a tradition can locate its texts in non-human agency, or in no agency at all. In the first case, the texts will represent, say, the voice of God; in the second, the texts will be *authorless*: the texts derive their authority precisely from their *authorlessness*. While the first, divine option has its own difficulties, the second option is theoretically interesting. All it needs is to come up with a conception of language, which is, in some sense, speaker-independent.

I think this is the source of the concept of *apauruṣeya*—impersonal words, authorless—in the Indian tradition; it may have been the rationale, as well, for centuries of vigorous discussion on language in this tradition. Texts, in the tradition, literally fell from the sky. In the ancient *Rgveda* (c. 1000 B. C.), for instance, the phenomenon of language is once described as a 'spirit descending and embodying itself in phenomena, assuming various guises and disclosing its real nature to the sensitive soul' (4.58.3, Coward 1980, vii). Since there is no (human) author, there is no preferred point of view; the basic texts of the tradition, the *Vedas*, are *svatah pramāna*, self-evident. Hence, the text cannot be questioned. In a way, then, it would seem that the vigorous and intellectually satisfying discussion of language— which marked the high culture of the tradition, as we saw—was *responsible* for inhibiting the emergence of the concept of knowledge and, therefore, of modern science.

I have been more than hinting in the last few paragraphs that I consider the Indian tradition as one where the concept of ascriptive knowledge need not be entertained. This is an empirical and archival issue beyond the scope of this chapter. Nevertheless, it is unlikely that such a vast and complex tradition can be fully captured in the simplified model sketched above. Also, there are bound to be serious exceptions to the model in the actual intellectual enterprise pursued over several millennia. Yet the model does seem to capture some of the crucial problematic features of the tradition; it seems to provide some basis for asking deeper empirical questions about the tradition.

References

Coward, H. 1980. *Sphota Theory of Language*. Delhi: Motilal Banarasidas.

Davidson, D. 1973. On the very idea of a conceptual scheme. In *Inquiries in Truth and Interpretation*. New York: Blackwell, 1984.

Davidson, D. 1975. Thought and Talk. In S. Guttenplan (Ed.) *Mind and Language*. Oxford: Clarendon Press.

Everett, D. 2005. Cultural constraints on grammar and cognition in Pirahã. *Current Anthropology*, 46, 621–646.

Gettier. E. 1963. Is justified true belief knowledge? *Analysis*, 23.

Hinzen, W., and M. Sheehan. 2013. *The Philosophy of Universal Grammar*. Oxford: Oxford University Press.

Kant, I. 1787. *Critique of Pure Reason*. Translated by N. K. Smith. London: Macmillan, 1929.

Mohanty, J.N. 1992. *Reason and Tradition in Indian Thought*. Oxford: Oxford University Press.

Mohanty, J.N. 2000. *Classical Indian Philosophy: An Introductory Text*. Lanham, MD: Rowman and Littlefield Publishers.

Mukherji, N. 2000. Traditions and the concept of knowledge. In *Science and Tradition*, A. Raina, B.N. Patnaik, and M. Chadha (Eds.). Shimla: IIAS.

Mukherji, N. 2002. Academic philosophy in India. *Economic and Political Weekly*, 37(10, March).

Mukherji, N. 2006. Beliefs and believers. *Journal of Philosophy*, Calcutta University, December.

Nagel, T. 1986. *The View From Nowhere*. New York: Oxford University Press.

Putnam, H. 1994. *Renewing Philosophy*. Cambridge: MIT Press.

Raju, C.K. 2007. *Cultural Foundations of Mathematics: The Nature of Mathematical Proof and the Transmission of the Calculus from India to Europe in the 16th C*. New Delhi: Pearson Education India.

Sen, A. 2005. *The Argumentative Indian*. London: Allen Lane.

Sen, P.K. 2007. Some problems of knowledge. In *Knowledge, Truth and Realism*. New Delhi: Indian Council of Philosophical Research.

Shah, N. 2003. How truth governs belief. *The Philosophical Review* 112(4): 447–482.

Chapter 9
Beliefs and Believers

> *There are aspects of higher mental processes into which the current armamentarium of computational models, theories, and experimental techniques offers vanishingly little insight.*
>
> Jerry Fodor

The concept of mind is an uncomfortable topic for contemporary philosophers. On the one hand, it is difficult to hold on to Cartesian dualism, at least in its received advocacy of mind and body as separate substances.[1] The difficulty was aggravated by Gilbert Ryle's 'deliberately abusive' characterization of the mind as a ghost in the machine (Ryle 1949, p. 17). On the other hand, it is hard to swallow the idea that the conscious, thinking subject is largely a myth. We may wonder if Ryle himself would have been comfortable with the view that his wonderful book, and the praise he received for it, were mere products of complex behavioural dispositions.

Given the fundamental dilemma, we find philosophers declaring that the mind–body problem is not a problem, it is a mystery (McGinn 1989). With respect to the more specific topic of beliefs as mental states, there are philosophers who have advocated 'no-theory' (Schiffer 1987). As with the discussion on consciousness earlier (*this volume*, Chap. 7), the suggested uncertainty regarding the concept of belief as mental states is my basic interest. I will focus on this interest by asking *afresh* what beliefs mean. In that sense, the discussion is not likely to appeal to the philosophers and cognitive scientists—some of whom will be discussed below—who have convinced themselves that the current semantic and psychological approaches on beliefs are fundamentally satisfactory.

This is a revised version of a paper published as Mukherji (2006).

---

[1]There are ways of making Descartes' distinction without substance dualism (Mukherji 2000, 2010a).

© Springer Nature Singapore Pte Ltd 2017
N. Mukherji, *Reflections on Human Inquiry*,
DOI 10.1007/978-981-10-5364-1_9

9.1   Beliefs and Mind

The concept of belief plays a rather curious role in the context of the fundamental dilemma. Beliefs are simply taken to be *mental states* with content. The postulation never really gets into the question of what *is* the mind, whether it is some part of the brain or something else. The approach is reasonable in the light of the history of sciences. Physics never really ventures directly into the question 'What is the physical universe?'; it studies what are taken to be the *states* of the physical universe. In studying the states of the universe, the physicist is led to study the constituents and the processes involved in those states, and a view of the universe gradually unfolds.

In philosophy of mind, the approach is bolstered by the observation that what are taken to be states of the mind—mental states—seem to have delineable content. It is thought that beliefs are couched in a certain form—known as *propositional form*—that indicates the content of the associated mental states. Donald Davidson (1975), for example, identified having of thoughts—and thus of minds—with the possession of these *propositional attitudes*. According to Davidson, animals do not have propositional attitudes since animals cannot 'talk'; therefore, animals do not have thoughts (and minds).

Thus, the English sentence, *Galileo believed that the Earth was flat*, is viewed as expressing a propositional attitude which ascribes a certain content to Galileo's belief, namely, that the Earth is flat. The suggested ascription, in effect, views Galileo's mental state in terms of the content of the embedded proposition that the Earth was flat. We will study these propositional contents below. For now, the point is that the strategy just outlined creates a nice zone of comfort for philosophers since it enables them to adopt the notion of mind to examine its contents in detail, while staying away from both Descartes and Ryle at the same time. Philosophers are perennially thankful to Gotlob Frege (1892) and Bertrand Russell (1919) for initiating this balanced move.

In fact, the parallel with physics works further. It is well known that physics might not have progressed so rapidly if the inquiry was devoted to general conceptual concerns about physical states, such as the distinction between physical and non-physical states, if any. Progress in physics was made possible by the identification of specific problems that can be addressed locally with definite empirical effect. The ongoing solution to these problems then formed the basis for understanding the general character of physical states. For example, the specific study of eclipses—among other things such as ebbs and tides—led to the postulation of the general design of the cosmos. In the case of study of the mind, philosophers have been engaged with some specific issues that seem to give intuitive and empirical handle on the otherwise elusive topic of the mind. These issues are concerned with propositional content.

As with the earlier discussion of phenomenal consciousness (*this volume*, Chap. 7) and concept of knowledge (*this volume*, Chap. 8), I will sketch these widely-discussed problems of propositional content in a rush. I will hardly engage with the immense, and fairly formidable, literature; as before, I will stick to some basic ideas and classic statements. My broad, hopefully valid, understanding of the subject is that, despite decades of voluminous discussion, no tangible solution to these problems is in sight. In this light, my strategy is to set these problems aside to shift to my own way of thinking about beliefs. To anticipate, the interest is that whether, on par with consciousness and knowledge, the concept of belief also may not be viewed as a mentalistic concept, at least not primarily.

The basic problem on the content of beliefs was raised by Frege (1892). Frege observed that, ordinarily, two co-referential terms of a language, such as *Cassius Clay* and *Muhammad Ali*, appear to be substitutable in a sentence without altering its truth-value. Thus, truth-value remains unaffected if we substitute the latter for the former in the sentence *Cassius Clay was a great athlete*. However, it is possible that the sentences *Narayan believes that Cassius Clay was a great athlete*, and *Narayan believes that Muhammad Ali was a great athlete*, differ in truth-value because, depending on what Narayan knows, Narayan may believe one but not the other. It appears as though the contents of Narayan's beliefs are (largely) unaffected by the fact that, in the world, Muhammad Ali is identical to Cassius Clay. Not surprisingly, Plato's allegory of the cave (*this volume*, Chaps. 2 and 8) springs to mind, as does Cartesian dualism, without any explicit mention of either.

To address the problem, Frege appealed to properties of language, rather than to psychological properties such as mental images. For what he called *proper names*, such as *the square root of four*, Frege held that there are two components of significance: sense and reference. It is commonly held that by *sense* of a proper name Frege meant something like the descriptive, conceptual content of the term 'in the head'; in that sense, the sense of a name is a constituent of *thought*, rather than of the world. By *reference* of a proper name, in contrast, he meant something like drawing attention to 'external' objects in the world. Since there are 'fictional' names such as *Sherlock Holmes*, it is held that the sense of a proper name may not always 'determine' its reference, even though sentences with fictional names, such as *Sherlock Holmes was an opium-smoking sleuth*, may be meaningful. I set aside the vexing problem that the sentence just cited could in fact be *true*, as Strawson (1952) would suggest: since (the character of) Sherlock Holmes was created by Arthur Conan Doyle, the cited sentence happens to be true in *his story*.

So, the solution to the problem of co-referentiality is that, in the case of beliefs, the proper name(s) in the embedded clause has only 'oblique' reference, that is, the embedded name refers to the sense of the term rather than to its 'customary' reference'. Since the names *the Morning Star* and *the Evening Star* differ in their conceptual content (sense), the substitution of one for the other in a belief-context—such as *Narayan believes that the Morning Star is red*—may fail to preserve truth-value even though the Morning Star = the Evening Star = Venus.

Although, as noted, it is doubtful if Frege himself explicitly held a mentalist view of beliefs, the subsequent literature found much basis for holding so (Linsky 1983). First, mental states are generally viewed as 'internal' to the subject such that their content may happen to vary from 'externalist' information (Fodor 1975); in fact, on occasion, they may be entirely fictional. In other words, beliefs seem to have *narrow* content restricted to the 'head.' Frege's idea that a proper name in a belief-context contributes only its sense to the computation of the meaning of the sentence thus reinforces the 'internalist' conception of beliefs. Second, we saw that the failure of truth-preserving substitution of co-referential names may be traced to Narayan's ignorance about how the world is like. Thus, the failure of substitution may be correctly ascribed to *Narayan*'s mental state, rather than to Kumar's, since Kumar may well know the relevant identity. In other words, content of beliefs may vary from one individual to another for the same item in the world. This consequence reinforces the idea that mental states are constituents of individual heads.

The trouble is that almost every aspect of the comforting picture just sketched can be challenged. Unsurprisingly, a very strong objection was raised in the philosophy of language. Saul Kripke (1972) argued that ordinary proper names such as *Muhammad Ali* are rigid designators such that the same name may be used counterfactually in different possible circumstances: *Muhammad Ali might not have been a Muslim, Muhammad Ali might not have been a champion boxer, Muhammad Ali might not have been an African-American*, and the like. Since the name has been meaningfully used to deny its apparent conceptual content (sense), we may conclude that ordinary proper names have no Fregean sense; Kripke also thought that proper names are not equivalent to Russellian descriptions. If that is so, proper names cannot have an oblique reference in belief-contexts because such reference does not exist. The most we can say about the name *Muhammad Ali* is that it 'picks out' Muhammad Ali, that very individual (in the world).

Yet somehow, despite the alleged absence of Fregean sense, the content of *Muhammad Ali* appears to contribute to the computation of truth-value for the sentence *Narayan believes that Muhammad Ali was a great athlete*; in fact, the problem is aggravated because the sentence *Narayan believes that Cassius Clay was not a great athlete* also admits of (possibly the same) truth-value (Kripke 1979). So, here we get a first glimpse of the possibility that the issue of significance of belief ascriptions may be unrelated to the issue of the content of the singular terms such ascriptions contain.

A range of philosophers (Donnellan 1972; Almog 1986; Wettstein 1981) extended the 'externalist' analysis of proper names to some definite descriptions as well: the so-called *rigid descriptions* as in *The man over there is drinking wine*. David Kaplan (1989) showed in detail that certain singular terms such as indexicals and demonstratives—*that, here, you*, etc.—are 'directly referential' in context; the content of these terms are allegedly totally determined by the (singular) object in hand. I am setting many controversies aside. But notice that each of these singular terms may occur as constituents of embedded clauses in a belief-ascription.

Perhaps more directly for the belief contexts at issue, Hilary Putnam (1975) suggested that even some common nouns, especially those designating 'natural kinds' like *tiger* and *electron*, seem to have properties of rigid designation.[2] For example, most native users of kind terms such as *elm* and *beech* cannot tell the

---

[2]I am setting aside another important problem that assimilation of a variety of syntactically distinct items of language—proper names, common nouns, indexicals, definite descriptions, etc.—under the same semantic picture of direct reference is implausible in view of design of language (Mukherji 1996).

conceptual difference between these terms. But these are not viewed by the users as synonymous because the users know that they designate different kinds of plants. That is why, in case of a dispute around such terms, we seek expert guidance on how things are in the world: *gold* is a telling example.

The upshot is that the (knowledge of) meaning of these terms, as they are used in forming thoughts, have an irreducible 'externalist' component along with whatever conceptual and encyclopedic content they have: these terms have *broad* content. To have a belief about elms, therefore, is to employ the broad content. From different directions, a range of authors (Salmon 1986; Bilgrami 1992) then showed how to construe beliefs to explain the presence of broad content; Salmon in fact claimed that, viewing singular terms as rigid designators, co-referential singular terms are substitutable even in belief contexts. Other authors attempted to combine Fregean and Kripkean insights by re-introducing Frege's notion of mode of presentation (Schiffer 1992; Fodor 1998). I set these more involved proposals aside (see Bach 1997 for a lucid review).

These developments had an uncertain effect on Frege's problem. I think a brief look at Jerry Fodor's views in the philosophy of mind shows why the scene is uncertain. Fodor had defended the narrow-content view of propositional attitudes for decades, arguing that the idea of broad content does not mesh with the idea of beliefs as mental states (Fodor 1987), as we saw. In the next section, I will discuss Fodor's idea of 'language of thought' in the context of folk psychology to examine whether ascription of beliefs amounts to identification of mental states at all. For now, it is interesting that Fodor ultimately agreed with the 'externalist' contentions of Putnam and other philosophers that common nouns designating concepts may have broad content as well (Fodor 1994; Fodor and LePore 1994); this puts pressure on the original Fregean notion of sense. Even later Fodor (1998) argued that the very 'internalist' idea of concepts is problematic; roughly, all we can say about the concept "dog" is that *dog* denotes dog. Setting much technical detail aside, suppose that the notions of rigid designation, direct reference and denotation form a family, insofar as the semantic value is concerned: thus, *John* rigidly designates John, *John* directly refers to John, *John* denotes John.

But then it could be argued that the notion of denotation is vacuous. Suppose, by *denotes* we mean something like 'stands-for'. Russell (1919) held that a symbol stands for something. Hence, the knowledge that *John* is a symbol is the knowledge that *John* stands for something. This general knowledge and the device of disquotation yields '*John* stands-for/denotes John.' On this view, the device of disquotation spells out that *John* is a symbol. One may know this without knowing what *John* (specifically) stands for or denotes; all we know is that *John* has (genuine) external significance, if at all. *If at all* because nothing prevents the notion of external significance to apply, say, to numbers including imaginary numbers, mental particulars, and other stuff that hardly qualify as items in the world. The argument extends to common nouns like *mountain*: one may know '*mountain* denotes mountain' without knowing what *mountain* means (Mukherji 2010b, Chap. 3). 'It is possible,' Chomsky suspects, 'that natural language has only syntax and pragmatics' (Chomsky 2000, p. 132); no semantics.

The upshot is this. Supposing the concept of belief to indicate mental states, the issue is whether the content of belief-ascriptions may be viewed as contents of mental states. It is natural to think of mental states as internal to the subject. If ascribed beliefs are to indicate mental states, we expect the contents of ascribed beliefs to have narrow content in the head, in the Fregean sense. In other words, it is natural to view the contents of mental states in conceptual terms in the language of thought, rather than in referential terms related to the world. Classically, mental states were viewed as hosting images, ideas, and other representational stuff; the internalist view of mental states attempts to capture the same domain somewhat formally in terms of the Fregean content of the constituents of thought. But it turns out that content terms, both singular and general, have predominantly broad content. So, propositional content does not match the desired content of beliefs.

In any case, despite much psychological investigations in the character of concepts (Murphy 2002; Carey 2009),it is unclear if concepts are genuine objects of inquiry; it is hard to find identity conditions for individual concepts without begging questions (Fodor 1998; Mukherji 2010b, Chap. 4). Thus, insofar as propositional content is concerned, all we are left with, if anything, are the objects 'denoted' by the terms of a language. Can the content of mental states be composed of 'denoted' objects? Apart from the near-vacuousness of the notion of denotation, it is totally unclear what it means for an object in the world itself to be the significance of a term. Notwithstanding causal and other purported accounts involving 'modes of presentation,' the problem aggravates if direct reference is supposed to supply an account of content of propositional constituents *in the mind*. No doubt when Kartik said *Hanuman carried the mountain*, Kartik had some or other mountain in his mind. But a theory of mind is supposed to unpack the familiar metaphor, rather than turn it into literal use.[3]

Notice that the *impasse* sketched above seems to arise when we assume the ascribed concept of belief to be a mentalistic concept. The brief history of the problem raises doubts about the genuineness of that assumption. It could be that a great deal of philosophical investigation since Frege had weak foundations insofar as the study of the mind was concerned; it now seems that we might have been wrong about what the ordinary concept of belief in fact is. Maybe the problem disappears once we locate the right ordinary concept of belief.

9.2   Troubles with Folk Psychology

Folk psychology develops the idea that the human mind is basically a repository of beliefs, lots of them, such that regularity in human action is explained via the regularity of beliefs that guide these actions. We explain why Mary came running

---

[3]It could be that all that there is to rigid designation, direct reference, and the like, is that content-terms of a language may have a variety of *uses* rather than a variety of meanings (Strawson 1950, 1961; Donnellan 1966; Mukherji 1989, 1995, 1996).

out of the building by ascribing to Mary the belief that the building is on fire. We ascribe many other beliefs to Mary to show that when these beliefs interact with Mary's belief that the building is on fire, and with Mary's basic desire to be safe, Mary's action of running out of the building follows.

Depending on the initial stimulus, some of the beliefs that interact with Mary's belief that the building is on fire, and with Mary's basic desire, may be listed as follows: that smoke is coming out of the window, that where there is smoke there is fire, that it is unsafe to continue to stay in the building, that it is safe to run out of the building. These, among others, will ensure that Mary gets up from her seat. Many more beliefs are needed to take Mary from her seat to the final exit. These will include, *inter alia*: that there is an exit, that she has an access to the exit, that it wasn't a case of false alarm, that the exit is to the front of the building, that the passage that leads to the front turns right at the next corner, and so on.

All of this and much else besides must have acted in an inferential chain to cause Mary to run out of the building. If the stimulus varies—for example, if someone shouted *fire*, or if the alarm system sets off—then a somewhat longer story is needed. If Mary is to act on a linguistic signal such as *fire*, then we need to ascribe various linguistic beliefs including beliefs regarding Gricean implicatures. Once these longer stories reach the point where Mary is picking up her bag, they rejoin the earlier story. In this way, an account of Mary's actions is given which generalizes over varying circumstances, capturing thus the regularity of Mary's beliefs.

One thing that follows immediately is that Mary's actions, in those fateful moments, couldn't have been the result of conscious, deliberate planning. Although she might have pondered briefly on her available options, she didn't have the time either to judge the entirety of her relevant beliefs, or to count them to see if she had enough of them, or to arrange them in a hierarchy to get the more particular ones from the more general. Basically, Mary couldn't have been introspecting about her beliefs at all. Since things happened so fast, her beliefs and the inferences reached from them must have been, for all that we know, *unconscious*. One could even argue that this unconscious 'modular' character of her beliefs ultimately saved her (Fodor 1983), unlike Buridan's donkey, which died pondering its options.

This consequence follows more naturally if we view Mary's mind to be essentially a moderately endowed digital computer with a complex programme already inscribed. As soon as some 'data'—namely, the smell of smoke—enter the system, computation begins rapidly. As harassed students of logic would testify, drawing a valid conclusion from even three premises, couched in the notation of quantification, is hard work. This is one reason why folk psychology found a natural host in digital computers. When the inferential needs of folk psychology was satisfied by the architecture of digital computers, a grand philosophical view emerged called *Turing Machine Functionalism*, or just *Functionalism*. As we know, even if the more generous philosophers of mind ascribe a (thinking) mind to computers, no one to my knowledge, with the probable exception of Chalmers (1996), ascribes consciousness to them (*this volume*, Chap. 7). The more unconscious Mary's computational mind, the more effective and adaptive it is (Pylyshyn 1984).

Beliefs, then, as folk psychology views them, are denizens of the dark, to use a phrase popularized by Quine. In this, folk psychology has at least two illustrious ancestors. Freud used the notion of *hidden* beliefs to explain various cases of psychosis—cases in which patients were unaware of the beliefs they held until Freud intervened. Also, the Socratic method essentially consists in digging up deeply buried salient beliefs that reside inside people. Socrates himself was said to have suggested nothing: he only helped in getting some of the denizens out of the cave. With such ancestry, it is no wonder that the notion of unconscious beliefs typing mental states gained wide currency.

Postulation of Mary's psychological states via ascriptions of beliefs to her supplies a rather useful answer to the issue of how these psychological states are to be described. Since, according to the view just sketched, an attribution of a particular belief entails that the attributee is in a psychological state, it is quite natural to think of these psychological states as individuated just as the particular attributions are individuated. So when I (truthfully) say *Mary believes that the building is on fire*, her psychological state must be individuated just as the relevant token of the embedded clause *the building is on fire* is individuated.

As we saw, a token of *the building is on fire* is individuated in terms of its form and content, namely, the form and the content of the English sentence *the building is on fire* in the given occasion of its use. So, if we knew how to capture the form and the content of this English sentence—its syntax and its semantics—then we know how to capture the form and the content of the relevant psychological state. So, doing enough philosophy of language with some caveats about causal connections, token-identities etc., amounts to doing enough philosophy of mind. I have already discussed serious problems with determining the content of belief-sentences. Here I wish to sketch another problem: the problem of same-believers.

Sentences belong to particular languages, English for example. So if Mary believes that the building is on fire, and Shyamali believes that *bārithe āgoon legeche*, then Mary and Shyamali must be in different psychological states, even if bilingual speakers of English and Bangla certify that the two sentences are identical in meaning. Since this plainly clashes with our ordinary notion of belief, there must be a neutral language of thought available to both Mary and Shyamali to whose representations the cited sentences of English and Bengali have a unique translation.

Recall once again the things that need to converge for this story to go through. The attributer comes up with a sentence, the sentence attributes to the attributee a certain belief contained in the embedded clause, and the attribution of the belief via the embedded clause signals that the attributee is in a certain psychological state identified by the content of the embedded clause. The language of thought hypothesis requires and ensures that things do converge this way. So what causes Mary to run out of the building is some sentence in the language of thought, among other things, where that sentence itself is derived, fast and loose, from a very large number of other sentences in the language of thought.

As Jerry Fodor has argued over the years (1975, 1981, 1987, 1994, 1998), the language of thought hypothesis is forced on us because we are attracted to two theoretical ideas as we saw: the theory of mental representations, and the theory of computation following Turing. The hypothesis thus is a theory-driven idea and, is, therefore, no part of our *ordinary* concept of belief unless explicit argument is marshaled to show that the human community is a community of functionalist philosophers. Pending that argument, the hypothesis needs to be judged on independent grounds, that is, grounds other than the availability of our belief-ascribing sentences.

There is a significant body of literature which suggests that the hypothesis doesn't work because things just don't converge in the desired ways. Accounts of regularities of beliefs fail to predict regularities of actions; regularities of actions cannot always be traced back to regularities of beliefs. We ascribe differing beliefs where folk psychology makes no distinctions; we generalize, that is, ascribe identical beliefs where folk psychology requires distinctions. These objections by now seem overwhelming. One could get a feel of the general problem as follows.

Most discussion on propositional attitudes takes a rather 'metaphysical' view of the scene. For the sentence *Mary believes that the building is on fire*, the literature, with a few exceptions, assumes that the proposition expressed by the stated English sentence designates a state of affairs: there's Mary, and there's the propositional content of *the building is on fire*, and the predicate *believes* signals a 'relation' between the two (Barwise and Perry 1983). So, the cited sentence is viewed as being on a par with *Narayan is taller than Kartik* in which the individuals Narayan and Kartik have a taller-than relation. This explains why the philosophical discussion basically concerns the character of this relationship. Frankly, I have never understood what it means for Mary—the person herself—to be *related* to a proposition, but I will let it pass because I understand that people are sometimes wedded to their ideas.

It is seldom explicitly mentioned that the sentence *Mary believes that the building is on fire* signals an *ascription* of a certain belief to Mary. Someone ascribes this belief to Mary, and when the ascription is valid, the (ascriptional) sentence *Mary believes that the building is on fire* turns out to be true. My linguistic intuitions suggest that that is not the case with *Narayan is taller than Kartik*. Even though this sentence also needs to be said by someone—because sentences do not fall from the sky except in oracles—it does not look as though the speaker *ascribed* the taller-than relation between Narayan and Kartik. It is more natural to say that the speaker, on that circumstance of use, etc., *described* the relation between Narayan and Kartik (see *this volume*, Chap. 7). Therefore, before we enter the discussion of the truth-content of the sentence *Mary believes that the building is on fire*, the study of propositional attitudes requires clarity on what it means for someone to ascribe a belief to Mary.

Recall the original story. *We* saw Mary running out of the building. Then *we* explained Mary's frantic action by ascribing a range of beliefs to Mary, including the belief that the building was on fire. Notice, in the story, Mary did not say a thing. Then, just by looking at her running, how could I say what she believed? A natural thought is that I myself would have been guided by such beliefs had I been placed in Mary's situation. That is, my ascription of a belief to Mary, say, that the building is on fire, can go through only if, were I to be in Mary's shoes, I would come to have that belief. In other words, first I counterfactually ascribe a belief to myself in thinking that the building is on fire. Then I factually ascribe this very-same belief to Mary. Given the correct circumstances as I view them, Mary and I would exchange our shoes. Mary and I are believers of the *same type* via our being same-sayers in the language of thought (for *same-sayers*, see Davidson 1974).

But suppose our shoes don't fit. Suppose Mary belongs to a very different believer-type. Thus Mary rushed out of the building only to take some pictures of the building on fire before she rushed back in to take more pictures inside; she came out of that door not because it was the nearest exit, but because it offered the best view; Mary is an iPhone addict. Then, of course, many of my ascriptions will not go through. It is easy to construct even more problematic cases with 'different' people, such as those who suffer from, say, colour-blindness. And these are just next-door cases.

The result is that, for various categories of 'weird believers'—animals, children, people with acute psychosis, class enemies, people with brain damage, aliens, foreigners, authors of the past, postmodernists etc.—either we say, dogmatically, that my ascriptions *make* them have those beliefs simply because I could entertain them, or, giving an account of the psychological make-up of these folks falls outside the scope of science. I was compelled to include postmodernists in the list after I read that Alain Badiou (2016) believes that fascists suffer from 'dislocation of language' (discussed in Mukherji 2017). How can I track Badiou's beliefs?

What options are now available for the general concept of belief? Notwithstanding protests from traditional believers (Baker 1987), one answer is that we should give up on beliefs, and focus directly on the properties of psychological states themselves. In other words, the suggestion is to de-link the theory from philosophy of language, and view psychological states as properties of brain-states. The first step in that direction is to deflate the concept of belief from a real to a notional one (Dennett 1982); we keep to beliefs not because they are there, but because we need them. In the next step, we show that the concept of belief is just not required (Stich 1983; Churchland 1981). So, we keep to the idea that psychological states are typically unconscious and mechanical without routing this idea through beliefs.

We may be inclined to accept this form of explanation insofar as Mary's action of running out of the building on being singed by the fire is concerned. Almost every moderately complex organism, with or without iPhones, is likely to so behave in such circumstances, independently of whether we ascribe beliefs to them or not. We may not be so inclined insofar as Mary promptly walks out of the building a few seconds before the actual fire starts, as Mary herself is a party to the conspiracy. We await a neural story that describes conspiratorial brain states. Suppose there is no such story.

Personally, I am not particularly worried that neither folk psychology nor cognitive neuroscience gives an account of people's beliefs and actions with sufficient generality. My worry lies elsewhere. Recall that folk psychology supposedly generalizes over our ordinary concept of belief. So, if the folk psychological notion of belief is to be set aside as descriptively useless, so must be case with our ordinary concept of belief. I wouldn't be worrying if the folk psychological notion of belief was merely a technical notion and *that* is found to be useless. We do give up scientific terms without residue and regret—*ether*, for example. But here we are asked to give up the baby of ordinary belief itself.

This is somewhat hard to swallow. I do continue to say, on occasion, that Mary believes that *p*, and, in saying that, I do continue to ascribe beliefs to Mary. I do not see how I can disengage myself from such activity even if folk psychology has turned out to be false. Am I then persisting blindly under the influence of a false theory? It is quite possible that the folk psychological concept of belief, given its illustrious ancestry, surrounds my ordinary concept as well in part; so this part needs to be given up as folk psychology is given up. Yet, since folk psychology contains certain purely theory-driven ideas such as the notion of hidden beliefs, the language of thought hypothesis, etc., it is unlikely that our ordinary concept of belief, which I continue to find profitable to use, is *entirely* infected with the requirements of folk psychology.

However, this last speculation will not be of much theoretical value unless I am able to argue that not only is folk psychology wrong as above, it is also wrong at least in part about the concept of belief I use when I ascribe beliefs to Mary. In other words, I need to argue that, in the garb of extending the ordinary concept of belief, folk psychology is, in fact, using a technical concept of its own.

Recall for the last time that folk psychology needs two ideas crucially among other things: one, that in ascribing the belief that *p*, I am stating that the subject is in the psychological state that *p*; two, that both myself and Mary are believers of the same type. The trouble, as we saw, is that an account which is restricted to believers of the same type isn't much of a general account of the psychological states of individuals whose actions are supposed to be captured by the theory. Could it be that our ordinary concept of belief is such that it *is* supposed to be used by—and among—believers of the same type, and that it is *not* intended to be used to describe psychological states at all? Is this where folk psychology, and much of the philo-sophical tradition, puts words of its own in the ordinary mouth?

It goes without saying that when I ascribe the belief that *p* to Mary, Mary must be in some psychological state or other. But this fact about Mary's mental life may have nothing to do with the content of my ascription. Our ordinary concept of belief, in that case, will not serve the purposes of the psychologist or the func-tionalist philosopher of mind.

9.3   What Believers Believe

I wish to examine whether the preceding conclusion can be reached by looking closely at some of the cognate words, especially the word *believer*. The task here is a modest one. We already have some idea of the concept we are seeking, as detailed above. To that extent, if only to dissociate ourselves from folk psychology, we are led to another folk theory of beliefs, this time devoid of psychological content. If this theory is to get off the ground, there must be some cases of our actual usage of related terms in which the theory is displayed. If we can find those cases, then by examining the contours of such cases in detail, we might have a better hold on the folk theory of ascription I am anxious to promote.

The case I wish to look at in some detail concerns some typical usages of the word *believer*, or its opposite, *non-believer*. The study of beliefs leads to believers as follows. In saying *S believes that p*, I am ascribing a certain link between S and *p* in that S is thought to be a p-believer. So, in focusing on the term *believer*, I am not going out of the gambit of terms at issue. Similarly, if S does not believe that p, that is, if S *dis*believes that p, then S is a p-non-believer.

If it is already beginning to sound bad, so much the better. There is a fairly standard usage of *believer* or of *non-believer* in which the notion of a *p*-believer is not explicitly invoked. We say simply that Gandhi was a believer, or that Bertrand Russell was a non-believer, period, without explicitly mentioning the particular belief that Gandhi held, or the one that Russell did not hold. Given folk psychology, in which Gandhi's or Russell's mental lives are completely described by detailing their beliefs (and their connections to stimulus and behavior), it is hard to make literal sense of such ascriptions. Since Gandhi and Russell were rational agents, the ascription that Gandhi was a believer amounts to a truism, and the ascription that Russell was a non-believer amounts to a contradiction, strictly speaking. It just couldn't be the case that, given the set of all beliefs, Russell didn't believe any of them   (see Dennett 1981, note 1).

The ascriptions then are not to be taken in their literal senses. The ascriptions are to be taken to mean that while Gandhi believed in God (or Ram Rajya, or whatever), Russell didn't. Especially for Russell, the ascription is to be understood in the context of a particular belief that God exists, and Russell was said to have denied *that* and only that. No sweeping denial of Russell's rationality was ever intended.

Before we look at the case with more details, notice that I am using the locutions *believes in* and *believes that* more or less interchangeably. It is quite possible that big conceptual issues lie in distinguishing between the locutions. However, it is likely that in some cases at least the contents of the two locutions converge. So when I say that Gandhi believed in Ram-Rajya, I mean to say that Gandhi believed that Ram-Rajya was both desirable and attainable. Let us pretend that we keep only to such convergent cases.

Returning to Russell's disbelief, the issue is this: how could we, on simply hearing that Russell was a non-believer, jump into a rather specific belief? Why didn't we ask, before we made sense of the report, *Russell didn't believe in what*? Suppose someone reported that Russell was a non-knower or a non-perceiver or a non-rememberer or some such stuff which, we are taught, comes in a (mentalist) package with non-believer. In each such case, we are certain to raise our eyebrows and, if we are polite, we will at least ask, *Russell didn't know what?*, *Didn't perceive what*? Without further specifications, such reports verge on incredulity and are, therefore, not made. If we want to draw attention to Russell's ignorance, lack of perception, failure of memory etc., then we are likely to come up with particular reports: Russell didn't know Urdu, couldn't smell the difference between *jeera* and *kalonji*, etc. How do we get away with a non-particular report about Russell's beliefs to have a particular effect?

We may get out of this problem by suggesting that the usage of *non-believer* at issue is a quasi-technical usage of the term reflecting the oddities of English language and the culture in which the usage is embedded. Given the trauma of the alleged conflict between the Church and science, occidental culture became obsessed with classifying individuals, especially prominent individuals, in terms of whether or not they believed in God. The persons who did or did not believe in God surely did or did not believe in other things. That goes without saying and is of not much consequence. What is of consequence, for the culture at issue, is whether or not they believe in God. The use of *non-believer* enables us to get the focus because the belief concerned does stand out in the culture in which it is embedded.

The point of worry in this otherwise plausible explanation is the claim that here we are dealing with a quasi-technical usage of an ordinary term. The suggestion, to repeat, is that I have used *believes* in the standard way when I say *Mary believes that the building is on fire*. I have used *believer* in a quasi-technical, non-standard way when I say that Russell was a non-believer. But that is exactly what folk psychology would like to claim for itself. Is the explanation then already laden with a theory we are anxious to get rid of? If we are not in the grip of folk psychology, could we make a counter-claim to the effect that *this* use of *believer* is the standard one, or that it carries some of the central features of standard usage of *belief* and *believing*, while the folk psychological notion of belief is in fact the non-standard one? The difficulty is that, since the term *believes* or its cognates have been used to describe some aspects of both Russell and Mary, it is not at all clear how we choose between the opposing claims. Whose fact of language is it, anyway?

The trouble here seems to be that the same word *believes* or its cognates have been put to two rather different uses. Then, assuming that a unitary concept has been displayed in either case, we are unable to determine which concept it is. Still we have made some progress in the sense that we have been able to raise an alternative to the folk psychological concept of belief. That the same word or its cognates have been used in two varying contexts with indeterminate theoretical significance may be an oddity of the English language. So we may have a better theoretical hold on the issue if we can find some uses of words in which the alternative concept as found in the standard usage of *believer* is displayed without invoking the general term *belief* at all. This, as we noted, is not possible in English.

At this point, I find it instructive to look at the *Bengali* words *āstik* and *nāstik*. Notice the *in L*, that is, the *in Bengali*, clause here. I am willing to settle for Hindi as well, but I will definitely not settle for *in Sanskrit* precisely because I do not know Sanskrit, and I do not know what it is for a Sanskrit word to have an ordinary use now. These remarks are needed because it could well be the case that the Bengali and the Hindi words at issue originated, in some sense lost in history, from the Sanskrit words I am trying to stay away from. Yet it does not at all follow that the *etymology* of the Sanskrit words applies to the Bengali words as well. For example, it is said by the experts I consulted that the Sanskrit word *āstik* is derived from *asti*.

Since I do not know the meaning of the word *asti* as I do not know Sanskrit, I couldn't have been displaying my concept of *asti* in my occasional use of *āstik*. As far as I am concerned, the occurrence of *asti* in *āstik* is an orthographic accident, like, as Quine famously pointed out, the *nine* in *canine*. Ignorance does turn out to be a blessing on occasion.

Going straight to the Bengali words then, the Bengali–English dictionary I have states that *āstik* means a believer in God, in the scriptures etc.; *nāstik* means someone who does not believe in these things. As a competent speaker of Bengali, I agree with these statements and I also find, to my surprise, that identical statements have been made in the Hindi–English and the Sanskrit–English dictionaries I have. It does not, therefore, look like an accidental, colloquial feature of some dialect; *āstik* means believer-in-scriptures, period.

Now notice the meaning-giving statement for *āstik*. We can think of this statement as a definition of the word in which the concept of belief occurs only in the right-hand side; no concept of belief occurs in the left-hand side, that is, in the word itself. This contrasts sharply with the definition of *believer* in English, which means one who believes in God. The Bengali word then is immune from the difficulty which plagued the English word. Notice as well that, in the definition, the concept of belief occurs as a part of belief-in-God; an *āstik* is not just a believer: he is a God-believer or a scripture-believer. In English, in the usage under consideration, he is simply a believer.

It need not be the case at all, as we saw, that *believer* always means God-believer. It may well amount to a belief in the scriptures—may be even, not just *any* scriptures, but the *Vedas*—in a community different from the Judaic-Christian community in which a believer is taken to be a God-believer. The important thing is that, given a community with very definite cultural demands, certain beliefs stand out whose acceptance makes someone a believer. This point is well reflected in the definition of a *nāstik* which says that, apart from being a disbeliever, he is a heretic, an infidel.

Not much mileage about a concept should be drawn from a single usage of a word. It is the concept that is at issue, provided we get some usage in which the central features of a concept are displayed. What, then, are the central features of the concept of belief displayed in the (alternative) usages examined so far?

First, beliefs are held or rejected in the face of a community. As such, beliefs fall outside the scope of two basic pillars of a purported science of the mind, namely, individual psychology and methodological solipsism. Beliefs seek to mark certain general, community-wide practices held significant by the community. Apparently, the suggestion is covered by Quine's observation that *speech* itself is to be viewed as community-wide practices as we divide them in terms of our practical needs (Quine 1969), but let me set that generalization aside.

Second, in order for someone to be a believer or a disbeliever, one must be taken to *understand* the belief ascribed to him, and the community is assured in advance that he so understands. The typical evidence for such assurance is that the community has heard the attributee say so in the past. Only then the issue of whether someone is a believer or not arises. So one would not ascribe a belief or a disbelief to a dog, child, ancient author, an alien, and the like, since the community-wide competence is not currently available. A believer and a disbeliever agree on understanding, but disagree on acceptance. A belief then is an item of potential *dispute* between persons of largely similar mental make-up. That is why we are so disturbed in the presence of a disbeliever—so near, yet so far.

Third, attribution of beliefs, that is, saying that someone is a believer, has nothing to do with the psychological state of the attributee. Given a taxonomy of beliefs frequently debated in a community, we may withhold an ascription of any of them to some creatures, without claiming that the creature, say a child, is empty of mind. Further, it hardly makes any sense to *blame* someone—for example, by calling him *infidel*—if the problem with that person is that he lacks a psychological state. If anything, he deserves proper treatment, not blame. That is why we are so aghast with the belief that people can be treated out of their disbelief or that a belief can be literally injected.

9.4   Small Beliefs, Big Beliefs

Let us now see what light this concept of belief throws on an individual's repertoire of beliefs. For example, I wonder what I genuinely believe. Concerning the discipline of philosophy, for instance, I believe that philosophy need not be taught at the undergraduate level, and that study of Indian philosophy should occupy only a minor, but significant place in any graduate programme; further, that every student of philosophy must thoroughly master either generative linguistics or quantum physics, preferably both. These are some of my curricular beliefs. I entertain them with the full knowledge that most people who have largely similar mental make-up, namely, other philosophers in India, disagree with my beliefs. So, I am prepared to argue for them, find evidence to support my arguments, and I feel annoyed when contrary beliefs are simply implemented without arguments. In general, I am reluctant to give up these beliefs easily precisely because I had spent much reflective time on them.

But sometimes I am also uncertain about these beliefs. Their lack of acceptance in my community keeps me on my toes. Thus, I am not prepared to act fully on them without further acceptance in my community in the future. Let us say I have a *contemplative attitude* towards the subject matter to indicate the sum total of my attitudes towards philosophy curriculum. Similarly, when I ascribe beliefs to others, I hold that person to have a contemplative attitude towards a specific subject matter—that is, we expect the ascribee to hold a belief that is not generally shared by his community, that he is prepared to defend it though he keeps himself open to repudiation.

Notice that we typically ascribe beliefs to someone when we ourselves do not hold them. X typically ascribes a belief to Y to report to Z. The point of the report usually is that Y believes differently from X and Z: *he is past forty but he still believes in communism*. Sometimes, it is just the opposite. The point of the report is that Y also believes something that X and Z believe when the belief is not generally shared in their community: *Y believes that armed struggle is the only way*. And, of course, X may ascribe beliefs directly to Y to confront or agree with Y: *You believe in the parting of the Nile, don't you*? To give these a name, call them *Big Beliefs*. The point to note is that the application of the notion of a community is as flexible as that of big beliefs.

In contrast, I simply cannot locate any contemplative attitude, in the sense just sketched, towards the toothpaste I use. I know where it is, I know how much is left, I know the price, the taste, and so on. That is, thought of toothpastes occupies a fairly secured sector of my psychological space, and I act on them on a daily basis. These actions are not routed through a dispute, to convince someone who believes otherwise. Why stop at toothpastes? I know that Kant wrote the *Critique*, that Plato was a Greek philosopher, that Wittgenstein's *Tractatus* contains five basic propositions, that ice is cold, and fresh grass is green. I know a lot, and so did Mary, whose rushing out of the building started all this. Let us say, Mary and I have *adoptive* attitudes to these things. We may be mistaken in some cases, we will acknowledge our mistakes, and change our adoptions. As a contrast, call them *Small Beliefs*.

Before I proceed let me note briefly that the distinction between big and small beliefs is so obvious and significant that similar distinctions are sometimes mooted in the philosophy of mind literature, unsurprisingly. For example, Dennett (1981, p. 78) makes a distinction between 'humdrum' beliefs about the house one lives in, quality of sandals, etc. from scientific beliefs. Similarly, Sperber (1994, p. 55) suggests a divide between 'cultural' beliefs—for example, belief in supernatural beings—which are 'few' and which violate 'module-based expectation,' and the vast range of beliefs that satisfy 'modular expectations.' I will not discuss these schemes in detail because they match neither my classification of beliefs nor the explanatory use of it I am interested in. Dennett and Sperber announce their respective schemes as part of their psychological investigations on the mental make-up of people. Both classification schemes are too casual and vague to qualify as tools for psychological studies. Consider the belief that electrons are leptons. Is this a 'humdrum' belief in the community of particle physicists in Dennett's scheme? Or, is this a 'supernatural' belief in Sperber's scheme? It looks as though these schemes are reflections of Dennett's and Sperber's personal preferences, rather than psychological natural kinds.[4]

To return to small and big beliefs, various questions of detail and terminology arise at this point. For example, one may question my free use of the concept of knowledge in my description of adoptive attitudes. I did so to highlight the salience and truthfulness of small beliefs, in contrast to the built-in uncertainty of big beliefs. Much, almost all, of our small beliefs must be veridical in order for us to survive as rational agents (Davidson 1977). But then the point of the distinction between adoption and contemplation collapses if the concept of knowledge itself contains the concept of belief (but see *this volume*, Chap. 8). So let us give up the concept of knowledge, and settle for *cognition* instead for *small beliefs*, and reserve the notion of *belief* only for *big beliefs*. Thus I cognize various facts about my toothpaste,

---

[4]In particular, I find Sperber's idea of modularity of beliefs problematic (see Fodor 2000).

language, geometry, and geography, while I believe in the ultimate goodness of humanity.

The terminology just suggested seems to fit the general Indian tradition. Just as the ancient tradition did not seem to have the Western concept of knowledge as justified true belief, there was no equivalent (general) notion of belief as well. However, there is a rich array of concepts—*pramā*, *prāmāṇya*, etc.—approximating the notion of cognition (*this volume*, Chap. 8). Interestingly, the tradition also has a very prominent quasi-moral concept of *vishwās* which means any of trust, conviction, confidence or faith. Objects of *vishwās* include persons such as gurus, texts such as the *Veda*s, institutions such as *sanātan dharma*, and universal principles like goodness of life.

We may ask various questions of detail about how to distinguish between cognition and belief on a case-by-case basis. Can I, for example, have a contemplative attitude towards my toothpaste? It seems to me pointless to ask, with respect to this entertainment or that, whether a given entertainment is an item of cognition or of belief such that all possible entertainments neatly subdivide into two clean blocks, in advance so to speak. This is my basic disagreement with Dennett and Sperber above, as noted. All we need right now is a *synchronic* distinction between beliefs and cognition for each individual, shared in a complex network with other humans. In this conception, yes, there could be beliefs about toothpastes as well if there is a new brand that uses controversial ingredients, or if there is discovery that all toothpastes cause cancer. Beliefs, under the current conception, are going to be rare in any case: we really cannot afford to withhold judgement and action subject to further examination, and willingly seek opposition, except for a small number of items at a time. Host of other problems arise with dogmas, propaganda, and the like.

However, when we are armed with the proposed conception of beliefs, two consequences follow. First, it is now possible to be rather suspicious of the very notion of *hidden beliefs*. These are the denizens of the dark, we saw, which lie deeply buried in the person's unconscious without the person being aware of them, although he acts according to them. We saw that both Socrates and Freud, with differing motivations, required hidden beliefs. Socrates dug them up to show his patient the grandeur of the patient's thoughts; Freud dug them up to show the patient how miserable she is, and persuaded the patient to give them up.

There is something essentially sinister about these methods if we look at the actual things supposedly dug up. The line of treatment is not intended to help the patient remember where he kept his toothpaste, or how he can solve a given numerical problem; when those problems arise, people go to a medical professional specializing in neuroscience, not to the uptown shrink. The intention is to dig up *big* beliefs such as the nature of virtue, the true meaning of life, or anxieties regarding impending catastrophe. We have one way of saying now that beliefs are held with a contemplative attitude such that the believer is aware of them, is prepared to defend them, and that he entertains the possibility of a disbeliever. Beliefs are thus out in the open for everyone to see. This openness to community examination *is* what makes them beliefs.

The second consequence I wish to stress is that now radical scepticism is a rational possibility. It has often been argued that radical scepticism cannot be coherent in an interesting way since the sceptic must share a large set of beliefs even to formulate and announce his sceptical position (see *this volume*, Chap. 2). The set of beliefs he shares with the non-sceptic defeats the very motivation of being a sceptic who does not want to believe anything. So, the sceptic can at best be a friendly critic challenging this belief or that. Such critics abound in any case.

But given the notion that beliefs are few and open, and given the idea that any such belief is always held with a disbeliever in view, we can now entertain the idea of disbelieving everything. Beliefs, in this view, require a certain suspension of judgement in any case. What the sceptic does is to go all the way, and suspend every judgement. He actualizes the possible disbeliever in every case; we will always find him when we are seeking opposition in our contemplative attitude. This he does with all his cognitive abilities intact, while disengaging himself from the historical burden of beliefs that has been passed on to him. The distinction between cognition and belief enables the radical sceptic to step out of the skin of history, and turn into a philosopher.

References

Almog, J. 1986. Naming without necessity. *The Journal of Philosophy* 83(4): 210–242.

Bach, K. 1997. Do belief reports report beliefs? *Pacific Philosophical Quarterly* 78, 215–241.

Badiou, A. 2016. Talk on 9th November, 2016 at University of California, Los Angeles, co-sponsored by the program in Experimental Critical Theory and the Center for European and Russian Studies.

Baker, L.D. 1987. *Saving Beliefs*. Princeton, NJ: Princeton University Press.

Barwise, J., and J. Perry 1983. *Situations and Attitudes*. Cambridge: MIT Press.

Bilgrami, A. 1992. *Belief and Meaning: The Unity and Locality of Mental Content*. Oxford: Basil Blackwell.

Carey, S. 2009. *Origin of Concepts*. Oxford: Oxford University Press.

Chalmers, D. 1996. *The Conscious Mind: In Search of a Fundamental Theory*. Oxford University Press, Oxford.

Chomsky, N. 2000. *New Horizons in the Study of Language and Mind*. Cambridge: Cambridge University Press.

Churchland, P.M. 1981. Eliminative materialism and the propositional attitudes. *Journal of Philosophy* 78(2): 67–90.

Davidson, D. 1974. Belief and the basis of meaning. *Synthese* 27, 309–323.

Davidson, D. 1975. Thought and talk. In *Mind and Language*, S. Guttenplan (Ed.). Oxford: Oxford University Press.

Davidson, D. 1977. The method of truth in metaphysics. In *Inquiries into Truth and Interpretation*, D. Davidson (Ed.). Oxford: Basil Blackwell.

Dennett, D. 1981. True believers. In *The Intentional Stance*, D. Dennett (Ed.). Cambridge, MA: The MIT Press.

Dennett, D. 1982. Beyond belief. In *Thought and Object: Essays on Intentionality*, A. Woodfield (Ed.). New York: Oxford University Press.

Donnellan, K. 1966. Reference and definite descriptions. *The Philosophical Review* 75: 281–304.

Donnellan, K. 1972. Proper names and identifying descriptions. In *Semantics of Natural Language*, D. Davidson and G. Harman (Eds.). Dordrecht, Holland: D. Reidel Publishers.

Fodor, J. 1975. *The Language of Thought*. New York: Crowell.

Fodor, J. 1981. *RePresentations: Philosophical Essays on the Foundations of Cognitive Science*. Cambridge: MIT Press.

Fodor, J. 1983. *The Modularity of Mind*. Cambridge: MIT Press.

Fodor, J. 1987. *Psychosemantics*. Cambridge: MIT Press.

Fodor, J. 1994. *The Elm and the Expert*. Cambridge: MIT Press.

Fodor, J. 1998. *Concepts: Where Cognitive Science Went Wrong*. Oxford: Clarendon Press.

Fodor, J. 2000. *The Mind Doesn't Work that Way: The Scope and Limits of Computational Psychology*. Cambridge: MIT Press.

Fodor, J., and E. LePore. 1994. Why meaning probably isn't conceptual role. In *Mental Representation*, S. Stich and T. Warfield (Eds.), 142–156. London: Basil Blackwell.

Frege, G. 1892. On sense and reference. In *Translations from the Philosophical Writings of Gottlob Frege*, P. Geach, and M. Black (Eds.), 56–78. Oxford: Basil Blackwell (1966).

Kaplan, D. 1989. Demonstratives. In *Themes From Kaplan*, J. Almog, J. Perry, and H. Wettstein (Eds.), 565–614. New York: Oxford University Press.

Kripke, S. 1972. Naming and necessity. In *Semantics of Natural Language*, D. Davidson and G. Harman (Eds.). Dordrecht: D. Reidel. Reprinted with an expanded introduction. Cambridge: Harvard University Press, 1980.

Kripke, S. 1979. A puzzle about belief. In *Meaning and Use*, A. Margalit (Ed.), 239–275. Dordrecht: D. Reidel.

McGinn, C. 1989. Can we solve the mind–body problem? *Mind,* New Series, 98(391): 349–366.

Mukherji, N. 1989. Descriptions and group reference. *Journal of Indian Council of Philosophical Research* 6(3): 89–107.

Mukherji, N. 1995. Identification. In *Philosophy of P.F. Strawson*, P.K. Sen, and R.R. Verma (Eds.). New Delhi: Allied Publishers (includes replies by Strawson).

Mukherji, N. 1996. Loads on reference. In *Epistemology, Meaning and Metaphysics after Matilal*, A. Chakravarty (Ed.). Studies in Humanities and Social Science, Special Number, Winter (Reviewed by Mark Siderits. 1998. *Philosophy East and West* 48(3): 503–513).

Mukherji, N. 2000. *Cartesian Mind: Reflections on Language and Music*. Shimla: Indian Institute of Advanced Study.

Mukherji, N. 2006. Beliefs and believers. *Journal of Philosophy*, Calcutta University, December.

Mukherji, N. 2010a. Doctrinal dualism. In *Materialism and Immaterialism in India and the West, Volume XII, Part 5 of History of Science, Philosophy and Culture in Indian Civilization*, P. Ghosh (Ed.). New Delhi: Centre for Studies in Civilizations.

Mukherji, N. 2010b. *The Primacy of Grammar*. Cambridge: MIT Press.

Mukherji, N. 2017. Everybody Loves a Good Fascist. *Revue des femmes philosophes*, January.

Murphy, G. 2002. *The Big Book of Concepts*. Cambridge: MIT Press.

Putnam, H. 1975. The Meaning of 'Meaning'. In K. Gunderson (Ed.), *Language, Mind and Knowledge*. Minnesota Studies in the Philosophy of Science, 7. Minneapolis: University of Michigan Press.

Pylyshyn, Z. 1984. *Computation and Cognition*. Cambridge: MIT Press.

Quine, W. 1969. Reply to Chomsky. In *Words and Objections*, D. Davidson, and J. Hintikka (Eds.), 302–311. Dordrecht: D. Reidel.

Ryle, G. 1949. *The Concept of Mind*. Chicago: University of Chicago Press.

Russell, B. 1919. The philosophy of logical atomism. *Monist* (Reprinted in R. Marsh (Ed.). 1956. *Logic and Knowledge*, 177–281. London: George Allen and Unwin).

Schiffer, S. 1987. *Remnants of Meaning*. Cambridge: MIT Press.

Schiffer, S. 1992. Belief ascription. *The Journal of Philosophy*, 89(10): 499–521.

Sperber, D. 1994. The modularity of thought and the epidemiology of representations. In *Mapping The Mind: Domain Specificity in Cognition and Culture*, L.A. Hirschfeld, and S.A. Gelman (Eds.). Cambridge: Cambridge University Press.

Stich, S.P. 1983. *From Folk Psychology to Cognitive Science: The Case Against Belief*. Cambridge, MA: Bradford Books/MIT Press.

Strawson, P. 1950. On referring. *Mind* (July) (Reprinted in P. Strawson, 1971. *Logico-linguistic Papers*, 1–27. London: Methuen).

Strawson, P. 1952. *Introduction to Logical Theory*. London: Methuen.

Strawson, P. 1961. Singular terms and predication. *The Journal of Philosophy*, 58 (Reprinted in P. Strawson, *Logico-Linguistic Papers*, 53–74. London: Methuen).

Salmon, N. 1986. *Frege's Puzzle*. Cambridge: MIT Press.

Wettstein, H. 1981. Demonstrative reference and definite descriptions. *Philosophical Studies* 40: 241–257.

Chapter 10
Varieties of Interpretation

> *The messy matter of human life should not be distorted to fit the demands of an excessively simple theory.*
>
> Hilary Putnam

As Immanuel Kant taught us, very little of the world comes to us solely via sensory channels, so to speak. We interpret vagaries of nature, traffic signals, musical scores and performances, visual arts, speeches and writings, smiles and tears, gestures and attitudes, practices and symbols, aches and twinges, and so forth; each of these categories come in a bewildering variety of individual forms. Interpretive activities differ not only with respect to the objects, but with the features of interpreters as well—their age, gender, interests and preparations, cultural location, and the like. It is hard to expect that we can discern some general pattern in these activities.

This is not to deny that there could be a general pattern, or a network of patterns, involved in all these. In fact, there must be: human interpretive practices cannot fail to have a common cause that is entrenched in the human design. But a discovery of that common cause could well be the agenda for a final science, if at all; *if at all* because the achievement of such a science may well fall beyond the scope of human design.

In the meantime, it is natural to settle for some version of what I have called *reflective pluralism* (*this volume*, Chap. 1)—the idea that human interpretive practices differ as interpreters and objects of those practices differ. In some favourable cases, such as certain basic levels of human linguistic interpretations or visual interpretations, we may hope to reach a somewhat more general thesis that detects commonalities between, say, sundry linguistic interpretations. Even there the claim of generality may well be restricted to some 'basic' levels such as grammatical interpretation, and may fail to cover pragmatic or even semantic aspects of interpretations (*this volume*, Chap. 5; Mukherji 2010, Chaps. 3 and 4). For the rest of our interpretive practices, some forms of literary criticism, and some reflective enumeration of *forms of life* are all that we are likely to get. In sum, the

---

This is a revised version of a paper published as Mukherji (2003).

© Springer Nature Singapore Pte Ltd 2017
N. Mukherji, *Reflections on Human Inquiry*,
DOI 10.1007/978-981-10-5364-1_10

point so far is that there are reasons to be suspicious about even modest claim of generality for this area of human conduct.

10.1 Interpretive Strategies

The American philosopher Michael Krausz (1993, 2000) is well known for his interdisciplinary attempt to offer a general theory of human interpretive practices. In what follows, I wish to examine one aspect of Michael Krausz's work from the stated direction of whether *any* general claim can be made. The theme I wish to cover has a larger spread than what Krausz has discussed. However, a focus on Krausz's work is prudent because, as we will see, Krausz offers a general thesis on human interpretive activities that pays careful attention to some aspects of the diversity of those practices. In that, Krausz's thesis is perhaps the most disarming general thesis currently in circulation. Yet, I will argue, even this disarming generality fails.

Krausz opens his book *Limits of Rightness* (Krausz 2000, LR) with the question, 'Must there be a single right interpretation for such cultural entities as works of art, literature, music, or other cultural phenomena?' (LR, p. 1). The form of the question suggests that Krausz expects a general answer: yes, or no, or something in between. Krausz settles for the third option as follows. *Singularism* is the thesis that a given 'cultural entity' admits of exactly one interpretation; *multiplism* is the thesis that cultural entities admit of more than one interpretation. Armed with these theses, Krausz concludes that, while singularism applies in some cases, multiplism applies in some others. Some entities, however, seem to escape this dichotomy when, in considering whether they admit of several interpretations, we are unsure if the same entity is involved in each of the allegedly competing interpretations.

Notice that Krausz's thesis is at once general and disarming. It is disarming because it denies rigid application of the notions of singularism and multiplism to cultural entities. Singularists, a rare species anyway, insist that every interpretable object admits of exactly one interpretation because there is a *unique human* point of view. On this non-relativist view, occasional appearance of more than one interpretation is illusory, and is likely to be a product of ignorance and inadequate reflection. Multiplists, in contrast, insist that human perspectives vary as cultures, traditions and practices vary; therefore, every cultural object must be open to a variety of interpretations. Krausz's suggestion is disarming because he admits *both* the possibilities; yet, it is general because Krausz think that the disarming possibilities obtain for every class of cultural entities. Obviously, the general thesis will be valid only if, while cultural entities vary, the notion of interpretation, that allows both singularism and multiplism for each sort, itself remains invariant. That is, if the notion of interpretation itself varies across classes of cultural objects, we will not know what sense to make of Krausz's generality claim.

Several points need to be noted to get some bearing on what these ideas mean. First, I presented Krausz's general thesis, following his opening remark, in terms of cultural entities alone. Soon we will see why. Krausz himself, however, wishes to extend the thesis to any object of interpretation whatsoever, cultural or non-cultural. I review this extension briefly below only to set it aside.

Second, Krausz's formulation of singularism and multiplism involves *admissibility* of interpretations, not just the availability of them. As I understand this interesting move, singularism and multiplism are to be viewed as the end results of a long and reflective interpretive process, not in terms of the beginning of this process. Consider the non-cultural example of the snake–rope problem widely discussed in classical Indian philosophy. You have the visual experience of some longish, greenish object lying in front you. 'Is this a snake or a rope?', you wonder. A given experience here gives rise to two possible interpretations. But this will not be an example of multiplism since both the interpretations have not been (simultaneously) admitted. In fact, in this case, both the interpretations cannot be admitted: just one of the interpretations can be right. That is why we proceed to inspect and admit one, if at all. Therefore, despite the availability of two possible interpretations, this at best is a case of singularism.

The proposed picture is as follows. We begin with, say, two available interpretations, and commence a process of investigation. First, we attempt to reject one or both interpretations. Suppose we are left with one; that is, singularism. Suppose we are unable to reject any. Then we try to put the two together to form a single coherent interpretation—called the *strategy of aggregating*. If the attempt succeeds, we get singularism once again in terms of the aggregated interpretation. If the attempt fails, we go back and try to 'pluralize' the original object so as to attach different interpretations to different objects. If the attempt succeeds, we get singularism once again. Otherwise, we get multiplism, where we are compelled to admit two opposing interpretations at the same time for the same cultural entity. Krausz is cautious to add that, here as well, we might prefer one of the interpretations over the other, although we can no longer explain our preference in terms of *rightness*. These processes, Krausz believes, apply to any object of interpretation.

I deliberately presented Krausz's proposals in an 'algorithmic' form to bring out the point that, on the face of it, the generality of his thesis concerns strategies of interpretations, rather than interpretations themselves. In other words, given that the evidence for objects of interpretations typically underdetermine the range of possible interpretations, we are likely to come up with more than one interpretation in most cases and proceed along the lines just suggested. This is just a methodological suggestion, which says nothing about the character of interpretations reached by this methodological route. For example, it does not prevent all interpretations to be exclusively singularist, or exclusively multiplist, or neither. That is, there is nothing in the description of possible choices that tell us how these choices are likely to be distributed. In that, Krausz's thesis by itself does not answer his leading question, 'Must there be a single right interpretation?'

It is not surprising that Krausz's thesis applies across the board, since we employ such strategies of rational inquiry in almost every sphere of possible dispute, as any ombudsman can tell. It is also not surprising that these strategies are immune from classical philosophical disputes around realism, since these disputes have to do with (the content of) interpretations themselves. In sum, making a list of logically possible options does not generate any substantive result. How do we add substance to Krausz's proposal?

For one, suppose Krausz suggested that, for any choice of object of interpretation, it is always the case that that object will give rise to all of singularism, multiplism, and neither. But Krausz does not say that, as we saw. More importantly, he cannot say that since, being oppositions, singularism and multiplism cannot apply to the same object when we fix the interpreter; multiplism obtains when all attempts at singularism fail. Alternatively, Krausz's proposals could have meant that, given the totality of all objects of interpretation, it is the case that some (not all) give rise to singularism, some to multiplism, and so on. This suggestion, though non-trivial, is far from being an interesting one. We saw already that the snake–rope case generates singularism at best. Classical figure–ground cases (duck–rabbit, face–vase, and so on) clearly generate multiplism. As we will see, Krausz himself holds, correctly in my view, that all music is multiplist. These are facts that immediately satisfy the alternative suggestion under discussion. But to say as much is to list some well-known facts; it does not say how these facts cohere. To appreciate the inadequacy of Krausz's view at this stage, notice that both the snake–rope and the figure–ground cases fall under visual experience, but the first satisfies singularism and the second generates multiplism. These are just broad facts, they don't furnish any substantive account of interpretations of visual experiences.

However, we may discern a more substantive contribution in Krausz's work when we take a closer look at the actual organization of his discussion. What follows, then, is a possible reconstruction of Krausz's work. First, I am interested in the fact that though Krausz had a general thesis concerning all objects in mind, he opens the discussion in terms of cultural entities such as literature, works of art, and music. Second, the greater part of Krausz's discussion on these issues over the years concerns cultural objects; there is only a marginal interest in non-cultural objects such as objects of scientific interpretation, middle-sized objects of common life, and the like. Third, even when Krausz ventures into non-cultural domains, he shows more interest in those cases, such as figure–ground cases, which have an intuitive pull toward cultural entities. Needless to say, these—in my view, selective—interests can be discerned much beyond Krausz to a variety of 'anti-essentialist' positions in recent decades. Such interests seem to show where Krausz's sympathy lies. How do we interpret the 'anti-essentialist' literature?

For obvious reasons, cultural objects—literature, marriage ceremonies, and religious practices—are generally viewed as grounds for the idea that interpretations vary as cultural locations vary. Hence, cultural objects naturally breed multiplicity. To cut a very long and confusing story short, this alleged fact of the multiplicity of cultural objects has led many authors in recent decades to announce multiplism, relativism, incommensurability, indeterminacy, and the like, for *all* human interpretive acts, including acts of scientific interpretation. A discussion of this turbulent literature is beyond the scope and interest of this chapter. In any case, I do not have much to add to Donald Davidson's devastating criticism of these dramatic announcements (Davidson 1973).

Krausz's contribution lies in distinguishing between multiplicity and multiplism even for cultural entities; just the availability of multiple interpretations is no ground for the suspension of rightness. That limit is reached when other options fail, as he rightly suggests. It is a confusion, therefore, to identify cultural entities in general with multiplism. In other words, the category of cultural entities—its ontology—has no direct links with the character of interpretive activities directed on them. Therefore, most philosophical theories, such as constructivism, constructive realism, and the like, that depend on a close tie between the category and its allegedly characteristic interpretation, are mistaken. This is not to suggest that these theories themselves are mistaken. They could well be right on independent metaphysical considerations (*this volume*, Chap. 2); but those considerations are now detached from the issue of interpretability of cultural entities.

However, Krausz may be viewed as challenging not only constructivists in their various guises, but realists as well. The argument consists of a resolution of two apparently conflicting claims. First, which is the central concern of this paper, Krausz maintains that 'multiplism is perhaps characteristic rather than definitive of the cultural' (*LR*, 12); yet he holds, second, 'multiplism is no criterion of the cultural' (*LR*, 11). The tension in these claims is difficult to miss. The only way I can interpret this set of puzzling claims is to think of a picture in which cases of multiplism and cultural entities cluster such that most cultural entities are cases of multiplism, and vice versa; yet cultural entities cannot be *identified* with multiplism since there will be, albeit rare, cases when some cultural entity satisfied singularism. To take an analogy, consider two oppositions: male–female and masculine–feminine. We could say that the female and the feminine cluster, though there are feminine males and masculine females. Thus, multiplism characterizes the cultural without being definitive of it.

In this picture, cultural multiplism is viewed as the central core of human interpretive activities. The rest of the activities trickle out of this core formation. Krausz is in agreement with the choice of the *domain* of the cultural relativist: this is where human interpretive activities are at their pluralist best. But Krausz is able to avoid the *position* of the relativist by allowing the core picture to diffuse at both ends: not only that singularism applies to some cultural entities, multiplism applies to some non-cultural entities. Thus, Krausz is able to disarm both the realist and the relativist, and any combination of the two such as constructivist-realist, while staying within the domain of cultural multiplism. The disarming subliminal message is that all interpretive activities are more or less cultural, but that concession does not by itself lead to any definite metaphysical position, say, about the necessary multiplicity of 'world-views'.

This, then, is the substantive consequence of Krausz's proposals regarding interpretive strategies. Given a conception of cultural entities (a category), the application of interpretive strategies to them generates both singularism and multiplism, though predominantly multiplism. For expository purposes, I am ignoring Krausz's third option of neither singularism nor multiplism. The availability of these strategic options enables a conception of non-cultural entities as a contrast to cultural entities. When the strategies are applied there in turn, we get both singularism and multiplism once again, but with the opposite distribution: singularism clusters with most non-cultural entities.

This is a very general picture of human interpretive activities that stretches across all domains of those activities, though with unequal weight, as we saw. We may view the earlier 'enlightenment' (modernist) perspective as advocating that the most salient form of interpretation is the scientific-determinist one giving rise to what Krausz will now call singularism; all other forms of interpretation were viewed as approximating the scientific one. With change in time and philosophical climate, there arose the opposite 'postmodernist' perspective in which every interpretation is relative to a variety of contextual hegemony such as traditions, ideologies, gender, ethnicity, and the like; basically, all human enterprises admit of multiplism, hence, indeterminacy. What Krausz does is to extract a more sobering friendly picture that places interpretable entities on all points of the entire spectrum stretching from rigid singularism to 'melting pot' multiplism.

These applications of interpretative strategies thus generate four broad areas: singularist–non-cultural, multiplist–non-cultural, singularist–cultural, and multiplist–cultural. The entire taxonomy, we saw, flows from the initial conception of cultural multiplism, the fourth area. Is that conception tenable? If the answer is in the negative, we need not even look at the other areas to examine the validity of Krausz's picture.

The crucial issue is that the conception of the category of cultural entities has to do with human interpretive activities in this area. In other words, it is assumed that there is something called cultural interpretation that applies to each of the entities that fall under this category. In turn, this means that there is some notion of interpretation that remains invariant across the entities in this category. To repeat, Krausz's programme requires, as with many other programmes in this area of philosophy, some coherent notion of culture that can be tied to the specific form of interpretive activity that takes place there. Traditionally, that tie had been sought in multiplism itself, namely, that cultural interpretations are distinguished by their abundance. But this is one of the 'orthodox' views that Krausz categorically rejects —correctly in my view. So, the only way Krausz can uphold a coherent notion of *cultural interpretation* is in terms of the nature of the interpretations themselves.

I will argue that the notion of interpretation varies so much across literature, painting, and especially music, that it is implausible—almost amounting to category mistakes—to think that unitary notions of singularism and multiplism apply everywhere. To take an analogy, consider Albert Einstein's claim that quantum theory is incomplete, and Kurt Gödel's claim that certain formal systems are incomplete. It does not follow that there is a general notion of incompleteness that applies to both the domains. In fact, if our concept of multiplism is understood in terms of the way it applies to literature, then it is hard to see that the twin concepts of singularism and multiplism apply to music at all. I will develop the argument from three different directions, and show that they converge.

10.2 Some Examples

To get a preliminary idea of the sort of problems I have in mind, let us consider some of Krausz's crucial examples that purportedly illustrate multiplism for different kinds of entities. Recall that Krausz's list of cultural entities includes other cultural phenomena besides literature, works of art, and music.

For general cultural multiplism, Krausz cites the interesting case of a dead baby floating in the Ganges river (*LR*, pp. 35–6). While Krausz himself, a North American, was plainly shocked by the sight of a 'dumped' baby, the locals explained to him that, being 'morally pure beings', dead babies are accorded the honour of being returned to the life source of the Ganges. The implication is that, while the North American Krausz was shocked, the locals would perhaps interpret the event as a holy gesture. Krausz asks: 'If we saw the same thing but interpreted it differently, who is right? Or is more than one interpretation admissible?' Plainly, what is involved here are large and irreconcilable belief systems of different cultures as embodied in their texts, convictions and practices: you cannot be both shocked (with disgust) and filled with religious admiration at the same time.

Suppose there is no doubt in this case that everyone sees a dead baby afloat. That is the sight. Then the question of whether the sight is repulsive or respectable depends on the cultural lens we use. Beginning with the visual experience then, there are two layers of interpretation: (a) the interpretation of a visual experience as that of a floating dead baby; and (b) the interpretation regarding how we evaluate (a). Multiplism in the sense concerned occurs, if at all, at (b). Thus, if the sequence of interpretations terminated with (a), multiplism will not apply. Moreover, even if there is prospect of multiplism at (a)—say, between a dead baby and a rotting idol shaped as a baby, the notions of interpretation involved here will be very different from the one applying at (b).

Next, consider Krausz's example of a 'pluralizing manoeuvre' regarding Vincent van Gogh's *Potato Eaters*. He suggests that this work of art may be subject to any one of formalist, psychoanalytic, Marxist, feminist, or other interpretations. To think of each of these interpretations as *cultural* interpretations, in the sense encountered in the previous example, raises several problems. For one, the formalist interpretation hardly involves any other belief system except the one solely geared for such artistic objects; for example, the formalist interpretation is likely to be concerned with tonality, strength of drawing, spatial arrangement, distribution of light, and the like. It will not apply to political systems, or family relations. In that, the formalist interpretation is very different in character from the other interpretations such as Marxist.

Suppose there is a unique formalist interpretation of *Potato Eaters*; so we do not yet have multiplism. The other non-formalist interpretations may be viewed as highlighting different aspects of the work such that, as Krausz suggests (*LR*, p. 13), it is possible to reach an aggregating interpretation, say, Marxist-feminist. The point is that it is hard to see that the admission of, say, a Marxist interpretation and a formalist interpretation will count as multiplism. Multiplism in this case can disappear in two different ways. This, as noted, cannot be the case at stage (b) of the previous example. What notion of a *cultural* entity then allows both examples to fall under the same category?

Consider now some examples from music. In the musical case, Krausz admits multiplism (2002; *RR*). But he also says that the notion of interpretation involved in musical multiplism attaches exclusively to performances of music, not to the scores themselves. In enforcing the restriction, Krausz is suggesting subliminally that the notion of interpretation in music might be significantly different from the interpretation of a work of literature. In the case of dramatic works, for example, there is a relevant notion of performance; hence, there is a relevant notion of interpretation in the sense of performance: actors interpret a play by acting it out in a particular way.

But a play also admits of varied cultural interpretations of the text itself: Marxist, feminist, Buddhist and the like. The point is too obvious to require illustration. Pieces of music, however, typically admit of only one of these forms of interpretation—as performance—even when we label pieces of music *romantic*, *baroque*, and so forth. This is not to deny that works of music admit of critical interpretations as well: popular, classical, muzak, mass music, war music, requiem, etc. Such critical interpretations, though distinct from performance interpretations, are closely tied to them. The present point is that even critical interpretations are very different from 'cultural' interpretations such as Marxist, feminist and so on. Music is a wholly different kind of cultural entity, if at all, than, say, a work of art or fiction.

The point can be illustrated by considering the factors that Krausz lists as contributing to multiplism in music. Starting with the idea that multiplism in music arises because 'musical scores are characteristically incomplete', Krausz suggests a number of 'different resources' in which 'different interpreters' interpret their music. These include choice of tempi, choice of timbre or volume of a given instrument, physical position of a musician in an orchestra, choice of bow movements for string instruments (up or down), duration and speed of a vibrato, pressures of bows and fingers, room temperature, well-accepted violations of the score, historical practices of a tradition, idiosyncrasies of a teacher, pressures on rehearsal time (*LR*, pp. 79–81), and the like. Cumulatively, Krausz calls these things 'extra-score practices'. Thus, his general conclusion is that, since 'extra-score practices vary historically', it will be incorrect to 'insist that the range of ideally admissible interpretations must always be singular' (*RR*, p. 87).

For the purposes of this chapter, I will not question whether such extra-score practices in fact lead genuinely to different interpretations. Let us assume so. Even then it is obvious that this notion of interpretation—hence, the related notion of multiplism—could not be the one that applies to *Potato Eaters* or floating babies. I will now attempt to give some theoretical shape to this concern.

10.3 Forms of Inquiry

It is an irony of human inquiry that, sometimes, different groups of people reflect apparently on the same object without having anything to say to one another. A classic example is the complete lack of conversation between astrophysicists and astrologers, though both deal with motions of stars. Astrologers think that stars have something to do with human fate; astrophysicists think that they are nothing but great balls of fire, totally incapable of influencing the course of psychic events.

In the star case, it is reasonably clear which inquiry is the valid one. In some cases, both inquiries could be equally valid, up to a point. Consider the distinction between theory of language and theory of literature. Both fields are concerned, in some sense, with the workings of language. Yet it is quite clear that they are looking at very different aspects of language and its use. A language theorist is basically concerned with a cognitive system; a literary theorist is concerned with a cultural-historical product with a cultural-historical content. Roughly the same holds for the more advanced forms of visual arts such as painting, sculpture and architecture. Thus, both literature and the visual arts may make comments, albeit in very different ways, on the futility of war, wickedness of power, grandeur of nature, personal grief, and so forth. One could conceive of an inquiry, which is focused exclusively on these comments and the explicit, articulated forms of making them. This inquiry need not concern itself with the properties of the cognitive systems of language and vision, which underlie the ability to make these comments. Nevertheless, we can also discern a difference between literature and the visual arts with regard to the distinction between cognitive structure and cultural comment.

For literature, it seems the distinction is overriding; it is hard to see how considerations from cognitive linguistics will bear upon the examination of the thoughts expressed in literature. The fundamental distinction seems to persist even when literary theorists self-consciously focus on 'linguistic' aspects of the literary work under discussion. An old classic example is Maurice Bowra's analysis of Coleridge's *The Ancient Mariner*. To bring out Coleridge's specific form of what he calls 'romantic imagination', Bowra examines Coleridge's use of language, including his poetic style, at great depth. *The Ancient Mariner*, Bowra suggests, 'draws attention to neglected or undiscovered truths' (1950, p. 68). The way a poet 'reveals' such 'secrets of the universe' is to 'work through myths' such as that of the ancient mariner. This myth is to be thought of as 'an extension of the use of symbols', where a symbol, according to Coleridge, is 'characterized by a translucence of the special in the individual'. In *The Ancient Mariner*, Coleridge 'shapes these symbols into a consistent whole' resulting in 'a myth about a dark and troubling crisis in the human soul'. Plainly, Bowra is concerned with 'world-views' as they get uniquely expressed in Coleridge's use of language. Although such concerns often require study of metaphor, irony, analogy and imagery, at no point do they require going into the structure of semantic interpretation, grammatical rules and pragmatic competence in ways in which linguists understand these things. In saying this, I am of course deliberately setting aside appeals to 'grammar', 'semiotics', 'illocutionary rules', and the like in much discussion of literature in the discipline of critical theory (see *this volume*, Chap. 11). I am not discussing this area because the notion of 'language' underlying this discipline does not concern the cognitive system.

No doubt some outstanding examples of art, such as Pablo Picasso's *Guernica*, Michelangelo's *Pietà*, van Gogh's *Cypress Tree*, the cave-paintings of Ajanta, and the like, are often understood in terms of their 'messages' on matters of human interest, as we saw for van Gogh's *Potato Eaters*. Yet, even in these exemplary cases, the predominant interest is in the form of the artistic piece rather than in its socio-historical content. In the large majority of artistic examples, especially for the more abstract and non-representational pieces, the interest is entirely in the form. The very fact that artistic pieces may be non-representational brings out the point under discussion. This point has little to do with the issue of realism in arts in the sense of 'realism' discussed in philosophical works, including this one (see *this volume*, Chaps. 2 and 3); a representational piece need not be realistic in the philosophical sense, as some of van Gogh's and most of Salvador Dali's paintings show.

The form of an artistic piece is intimately connected to how it appears to its viewers; that is, to its perceptual properties. In this sense, much of the visual arts may be thought of as skilled manipulation of perception. Hence, the distinction between cognitive system and the cultural object is less marked in our understanding and appreciation of the visual arts. In fact, there is by now a substantive literature on how great artists, such as Michelangelo and Leonardo da Vinci, exploited properties of the visual system—as it represents, say the distribution of light and shadows—to design their work. Cinema raises problems of classification that I will set aside in this discussion.

The intimate connection between study of the cognitive system and the resulting work of art led Ernst Gombrich (1960, p. 33) to cite John Constable with approval:

> Painting is a science and should be pursued as an inquiry into the laws of nature. Why, then, may not landscape painting be considered as a branch of natural philosophy, of which pictures are but the experiments?

The interest is that Constable's own work, as Gombrich puts it (34), 'is surely more like a photograph than the works of either a Cubist or a medieval artist'. Even then, as Gombrich's subsequent analysis of Constable's *Wivenhoe Park* brings out, 'the painter's experiments adjoin those of the physicists'. In this sense, the artist's achievement lies 'in the *discovery of appearances* that is really the discovery of the ambiguities of vision' (314). This is in stark contrast to Bowra's 'linguistic' investigations.

I am not suggesting that an artistic work, therefore, ought to be viewed on a par with the science of the relevant domain, say, the human visual system. The artist's scientific explorations, if any, at best underlie his artistic expression; unlike the scientist, he is not describing the visual system. In other words, although the artist's 'discovery of appearances' often requires some understanding of the concerned cognitive system, this understanding is exploited rather than expressed, much as advertising campaigns exploit the laws of human gullibility.

Nevertheless, it does follow, as Gombrich's extensive analysis shows, that one significant way of explaining a work of art is to explain the psychological understanding that goes into its making. In that sense, the distance between a psychological study of visual arts and their aesthetic study is not as far removed as it is for literature. In fact, aesthetic explanation is likely to converge onto psychological explanation at prominent joints. As we saw, the more a work of art is seen as a formal object—in contrast to a socio-historical product—the more amenable it is to psychological explanation. This raises the possibility that the distance between these forms of explanations becomes indistinguishable when a work of art is not viewed in terms of its message at all.

In the case of music, it is even more difficult to make a distinction between a cognitive system and a cultural product. Though no human enterprise can fail to be a cultural-historical product (and music is no exception), it is difficult to maintain that music has a cultural-historical content in that it makes comments on, say, the futility of war, although it is quite possible that the cultural-historical context of a war might have led someone, such as Igor Stravinsky, to compose a specific brand of music. This point about music can be brought out in several ways, as we will see.

For now, one quick piece of evidence is the widely-tested ability of very young children to intuitively grasp and perform fairly advanced forms of music even when they have very little 'world-knowledge' to grasp advanced forms of literature and the visual arts. This 'non-representational' character of music is a puzzle of great theoretical interest that raises doubts about the distinction between a cognitive system and a cultural product (Mukherji 2000). For this reason, it has been a persistent problem to incorporate music in aesthetic and critical approaches that begin with, say, architecture and Greek tragedy. It is also, I think, the underlying reason for an ancient interest in the 'language-likeness' of music that Leonard Bernstein (1976) finally and explicitly raised. It is interesting to note that the grammatical complexity of language (long-distance reflexives, triple embedding, double negation and so on) also poses no problem for young children even when they have troubles with metaphors, analogies, deliberate ambiguities, ironies, and the like.

In my view, current work in linguistics and musical cognition can be fruitfully linked to some of Wittgenstein's insights to develop the idea of 'language-likeness' in the grammatical sense just outlined. The perspective that ensues helps explain why musical interpretation is fundamentally different from literary interpretation. Musical interpretation, I will argue, stops at a level analogous to the grammatical level of interpretation (Mukherji 2010, Chap. 6 for more).

10.4 The Psychological Version

The suggested parallel between just the grammatical part of language and the whole of music can be approached from Ray Jackendoff's interesting discussion of these issues (Jackendoff 1992). He begins by distinguishing between two versions of 'the fundamental question for a theory of mind'. The philosophical version poses the question, 'What is the relationship of the mind to the world... such that our sentences can be true or false?' The psychological version poses the question: 'How does the brain function as a physical device, such that the world seems to us the way it does?' I am using Jackendoff's distinction largely for exegesis. There are problems with the formulation of his distinction. In particular, philosophical questions need not always be concerned with truth and falsity, and psychological questions need not make any mention of the brain or the world except insofar they are viewed as containing mental states. But these objections do not disturb the discussion in this section.

With this distinction in hand, Jackendoff argues, 'it hardly makes sense to say that the representations one constructs in response to hearing a performance of the *Eroica* are true or false' (Jackendoff 1992, p. 165). Mention of Beethoven's (later) work is particularly relevant here since Beethoven wrote the symphony 'in the absence of any overt musical signal'. Thus, it seems absurd to ask if the piece suddenly acquired a truth-value when the score was written or the first performance took place. The philosophical version of the fundamental question, therefore, does not apply to music at all; only the psychological version does.

It seems that the inapplicability of the philosophical version to music is nearly obvious. In his influential work, Roger Scruton (1983) has forcefully argued that musical symbolism does not imply that its symbols stand for anything in the world. Working through well-known examples of music where aspects of nature are allegedly depicted (blowing of wind, sound of waterfalls, bird calls, cries of animals), Scruton argued that no intelligible sense can be made of the idea that these sections of music either resemble or represent aspects of nature. Further, even if we grant that such music imitates nature in some way, we cannot say that the music *says* something about those aspects of nature. In sum, musical symbolism lacks predication in the desired sense. Since predication does not occur, there cannot be any (truth) satisfaction in Alfred Tarski's sense; hence, the notions of truth and falsity simply do not apply to music.

The interest here is that Jackendoff also makes a similar claim for aspects of linguistic research. Suppose language consists of three parts: grammar, phonology and semantics. Jackendoff claims that the psychological version holds, as against the philosophical version, for each of these parts. The claim is most controversial for the third of these parts; hence, Jackendoff's arguments for the psychological version of semantics are the weakest. There is a strong intuition that *dog* is true of dogs and, thus, the sentence *Dogs are feline* is false (Fodor 1998; Fodor and LePore 1994). Except for the general suggestion that terms such as *true* or *false* need to be 'embedded' 'in a general theory of concepts', Jackendoff has done nothing specific to dispel this intuition. I do not, thereby, mean to endorse the 'philosophical version' for semantics. My complaint is that Jackendoff's arguments for the psychological version in the domain of semantics are insufficient; the complaint extends to Jackendoff's own proposed semantic theories (Mukherji 2010, Chap. 4). In my opinion, Noam Chomsky presents a more powerful perspective in favor of the psychological version of semantics as a property of I (nternal)-language (Chomsky 2000; Mukherji 2010, Chap. 3).

However, Jackendoff's claim for phonological representations is more plausible since it is totally unclear what it means for the *noise* 'dog' to be true or false: it is 'difficult to see how the predicates *true* and *false* apply to one's phonological representations in response to an incoming stimuli' (164). Jackendoff concludes that 'the notion of computation need not have anything to do with "respecting semantic relations" at least in the domains of phonology and syntax' (29). Consider the aspirated sound *p*, as in *Patrick*. Jackendoff's central point is that these phonological objects themselves do not stand for something else. If you like, there is a phon-sound correlation between *p* and an aspirated sound; there is no further correlation between the sound and something else in the world. *Patrick* is just an arrangement of sounds. The point is even more compelling for objects in syntactic structures: 'There is no such thing as an NP, a VP, or an Adjective in the environment.' In sum, 'speakers do not believe (or believe in) NPs or phonological distinctive features or rules of aspiration' (165).

With these considerations in hand, it is worth asking if the significance of a piece of music—hence, its possible interpretations—ought to be phrased in any external terms, that is, terms that refer to elements apart from the musical symbolism itself, at all. Jackendoff shows that certain aspects of musical affect—for example, why certain pieces of music do not seem to lose their pleasing effects even after repeated hearing—can be explained solely in terms of the combinatorial properties of notes, and the modular character of musical processing (Lerdahl and Jackendoff 1983). The point is that it is possible to explain why we *want* to hear the same music from the properties of musical processing alone, not *because* the piece invokes—although it may—pictures of reality, desires, memories of first love, and the like.

Interestingly *and the like* is beginning to include even moods and emotions which are thought to be the hallmarks of musical significance in most common and philosophical conceptions of music. Granting that the external significance of music cannot be captured in terms of representations of aspects of reality, it is held by many that music represents emotions in a way that can be recognized by listeners. As Diana Raffman (1993, p. 42) cites Roger Scruton, it is 'one of the given facts of musical culture' that the hearing of music is 'the occasion for sympathy'. Thus, the literature on the emotional significance of music include: being merry, joyous, sad, pathetic, spiritual, lofty, dignified, dreamy, tender, or dramatic; feelings of utter hopelessness, foreboding, anxiety, terrified gesture, and the like. For Scruton, if someone finds the last movement of the *Jupiter Symphony* 'morose and life-negating', he would be wrong.

In recent years, cognitive theorists of music have generally rejected this tradition. The basic objection to the very idea can be stated in Raffman's terms as follows: 'Musicians argue about phrasings and dynamics and resolutions. They do *not* argue about the emotions they feel or otherwise ascribe to music' (59). As Raffman elaborates, musicians may argue that a given phrase ends at a certain E-natural because the note prepares a modulation to the dominant; the argument never takes the form that the note expresses ultimate joy, or whatever. I return to this example. None of this of course is meant to deny that listeners often have emotional responses to music. The point is that the fact need not be traced to music itself.

10.5   Wittgenstein on Language and Music

Ludwig Wittgenstein reached this point several decades before the onset of cognitive psychology of music. In his *Blue and Brown Books* (Wittgenstein 1958, BBB, 178), he remarked:

> It has sometimes been said that what music conveys to us are feelings of joyfulness, melancholy, triumph etc., etc., and what repels us in this account is that it seems to say that music is an instrument for producing in us sequences of feelings. To such an account we are tempted to reply 'Music conveys to us *itself*.'

According to him, 'it is a strange illusion that possesses us when we say "This tune says *something*," and it is as though I have to find *what* it says.' Given that what a tune says cannot be said in words, 'this would mean no more than saying "It expresses itself." To bring out the sense of a melody then 'is to whistle it in a particular way' (BBB, 166).

To see what Wittgenstein might have meant by his claim that music 'expresses itself', it is interesting to note that he extends the claim to language as well—to the understanding of a sentence, for example. He suggests that what we call 'understanding a sentence' has, in many cases, a much greater similarity to understanding a musical theme 'than we might be inclined to think'. The point is that we already know that understanding a musical theme cannot involve the making of 'pictures'. Now the suggested similarity between music and language is meant to promote a similar view of language as well, namely, no 'pictures' are made even in understanding a sentence. 'Understanding a sentence', he says, 'means getting hold of its content; and the content of the sentence is *in* the sentence' (BBB, 167).

There are several ways of interpreting these difficult claims. It is well known that Wittgenstein's own way is to draw attention to *gestalt* features of object-perception, which, in a way, leap into our minds. Hence, Wittgenstein devotes a major part of his analysis to properties of visual perception in an attempt to draw lessons to apply in turn to music and language. Even if we grant that lessons from vision might work for music, how can it work for language? For example, despite Wittgenstein's valiant attempts, it is hard to see how the notion of *expression*, as in 'what a face or a flower expresses', applies to what a sentence expresses.

In my opinion, the suggested parallel between understanding a sentence and a piece of music such that they convey themselves can be explained from an altogether different theoretical perspective. In this perspective, the significance of a sentence can be brought out in various layers, beginning with a layer that has no external significance at all. We can then view the other layers in terms of progressive addition of external significance. Each layer, nevertheless, admits of multiplism that attaches exclusively to that level. I will suggest that multiplism in music is very much like the multiplism of language at the initial level.

Consider the sentence *Who knows John gave what to whom*? The sentence admits multiple interpretations depending on the relative scopes of the embedded *wh*-phrases. Since these are questions, I have also included a possible answer in each case to display the differences of interpretation somewhat more perspicuously.

(i) Representation: who$_i$ e$_i$ knows to whom$_j$ [John gave what e$_j$]
   Interpretation: For which persons x and y, x knows John gave what to y
   Answer: Bill knows John gave what to Mary
(ii) Representation: who$_i$ e$_i$ knows what$_j$ [John gave e$_j$ to whom]
   Interpretation: For which person x and what thing y, x knows
   John gave y to whom
   Answer: Tom knows John gave the book to whom

Someone's knowledge of John's gifts is under query here. In (i), the query is about the recipient of those gifts; in (ii), the query concerns the gift-item. The sentence under discussion thus admits of multiplism. Representations (i) and (ii) are linguistic expressions *par excellence*—called *LF-representations* in linguistics (2010, Chap. 2 for fuller explanation). Hence, many aspects of the interpretations that can be attached to them are linguistic in character as well. In particular, we do not expect the shape of these expressions to be available in any other symbolic domain.

Nevertheless, I wish to draw attention to some general features of this example which, in my opinion, are available beyond language. First, (i) and (ii) are structurally distinct in that the relative positions of the symbolic objects in them differ. Second, these structural differences are directly related to the way a representation is to be interpreted. In fact, one of the global economy principles—called *full interpretation* in linguistics—stipulates that a representation may not contain any element that cannot be interpreted. Third, the interpretations do not make any reference to how the world is like, the beliefs of people interpreting them, the vagaries of the associated culture, and the like. In fact, in order to differ, the interpretations do not require that there be an external world at all. Yet, to emphasize, multiple interpretations are attached to the same symbolic object solely in terms of the ambiguity of its representational structure.

Consider again the possible dispute between musicians which Raffman suggested to show the irrelevance of emotivism for musical interpretations. The dispute concerned the identification of a musical phrase, whether it ends with a certain (occurrence of) E-natural. In principle, then, the dispute can be traced back exclusively to the structural features of how a group of notes are to be represented. Three possibilities arise: the phrase ends before the E-natural; the phrase ends at the E-natural; and the phrase extends beyond the E-natural. As anyone familiar with music knows, these structural differences make big differences in the interpretation of music. Depending on the group of notes at issue, and the location of the group in a passage, some of the structural decisions may even lead to bad music. This is because these decisions often make a difference as to how a given sequence of notes is to be resolved. Any moderately experienced listener to music can tell the differences phenomenologically, though its explicit explanation requires technical knowledge of music (such as modulation to the dominant).

This explains why composers and performers spend a lot of time marking a score to show how exactly they wish a sequence of notes to be grouped. Lerdahl and Jackendoff's work (Lerdahl and Jackendoff 1983) shows how different groupings impose different hierarchies on musical surfaces such that each hierarchical organization gets linked to a specific interpretation of the surface. The phenomenon is explicit in musical traditions that use scores. But it can be observed in any tradition by attending its training sessions, for example. Training means attention to the pitch of individual notes, *and* how notes are to be organized. When the music becomes complex, and it begins to tax memory and attention, various devices are used to highlight the salient properties of symbolic organization. These include emphasis typically by suitable ornamentation, organization of music in delineable cycles such as *rondo*, display of unity of larger sections by *cadence*s, exploiting the cyclic features of the accompanying beat, and so on. The list is obviously very incomplete, but it is pretty clear that, in some sense, there is nothing else to music. Interpretations in music are sensitive solely to the syntactic properties of representations.

Plainly, there is much else to linguistic interpretations. Consider Chomsky's example *Drinks will be served at five*. As Chomsky (1975, p. 65) observed, the sentence can be used as 'a promise, a prediction, a warning, a threat, a statement, or an invitation', among others. A decision about which of these varied interpretations of the given sentence is most salient will depend on the features of the extra-linguistic environment. These features include the states of mind of the speaker and her audience, a knowledge of the specific locale in which the sentence is uttered, some knowledge of the culture in which the given community of people generally participate, facial expressions, past utterances, and so forth. In sum, the linguistic object *drinks will be served at five* needs to interact with other systems of knowledge and belief for these interpretations to be available.

The array of these systems can get progressively thicker to include social relations, cultural choices, religious pronouncements, proto-scientific beliefs, conceptions of the future, and the like. At some point in such a dense field of interactions, we get works of literature. These works themselves can then seep into the general culture to generate even wider belief systems—most cultures are textual cultures, in that sense. Since we do not have the faintest idea of how these systems are organized with respect to each other, let us say that interpretations of literary and cultural texts form a continuum with items of common life such as *drinks will be served at five*. The entire continuum may now be viewed as distinct from syntax-governed interpretations outlined above. There is thus no general notion of interpretation that spans both literature and music, even if we want to place them under the common head *cultural entities*.

References

Bernstein, L. 1976. *The Unanswered Question*. Cambridge: Harvard University Press.
Bowra, M. 1950. *The Romantic Imagination*. London: Oxford University Press.
Chomsky, N. 1975. *Reflections on Language*. New York: Pantheon Press.
Chomsky, N. 2000. *New Horizons in the Study of Language and Mind*. Cambridge: Cambridge University Press.

Davidson, D. 1973. On the very idea of a conceptual scheme. *Proceedings and Addresses of the American Philosophical Association*, 47(1973–1974): 5–20. Reprinted in *Enquiries in Truth and Interpretation*. London: Blackwell, 1984.

Fodor, J. 1998. *Concepts: Where Cognitive Science Went Wrong*. Oxford: Clarendon Press.

Fodor, J., and E. LePore. 1994. Why meaning probably isn't conceptual role. In *Mental Representation*, S. Stich, and T. Warfield (Eds.), 142–156. London: Basil Blackwell.

Gombrich, E.H. 1960. *Art and Illusion: A Study in the Psychology of Pictorial Representation*. Princeton: Princeton University Press.

Jackendoff, R. 1992. *Languages of the Mind*. Cambridge: MIT Press.

Krausz, M. 1993. Rightness and reasons in musical interpretation (RR). In *The Interpretation of Music*, M. Krausz (Ed.). Oxford: Clarendon Press.

Krausz, M. 2000. *Limits of Rightness (LR)*. Lanham, MD: Rowman and Littlefield.

Krausz, M. 2002. Making music: beyond intentions. In *The Linacre Journal*, R. Harre, and J. Shosky (Eds.), 17–27. Oxford: Linacre College.

Lerdahl, F., and R. Jackendoff. 1983. *A Generative Theory of Tonal Music*. Cambridge: MIT Press.

Mukherji, N. 2000. *The Cartesian Mind: Reflections on Language and Music*. Shimla: Indian Institute of Advanced Study.

Mukherji, N. 2003. Is there a single notion of interpretation? In *Interpretation and its Objects: Studies in the Philosophy of Michael Krausz*, A. Deciu (Ed.). Amsterdam: Rodopi Publishers.

Mukherji, N. 2010. *The Primacy of Grammar*. Cambridge: MIT Press.

Raffman, D. 1993. *Language, Music, and Mind*. Cambridge: MIT Press.

Scruton, R. 1983. *The Aesthetic Understanding*. Manchester: Carcanet Press.

Wittgenstein, L. 1958. *The Blue and Brown Books (BBB)*, 1958. Oxford: Basil Blackwell.

Chapter 11
Literature and Common Life

> We learn much more of human interest, about how people think
> and feel, by reading novels or studying history than from all of
> naturalistic psychology.
>
> Noam Chomsky

The charming thing about the issue of convergence of literature and philosophy is that the issue is embedded in opposing intuitive pulls. On the one hand, there is a clear intuition that literature and philosophy, as forms of human thought, have significant convergence that goes deeper than the historical convergence of any pair of reflective forms of human thought. In other words, the intuition demands that the convergence of literature and philosophy be viewed in terms more intimate than what binds, say, literature and science, or literature and music. On the other hand, there is a strong intuition that literature and philosophy are very distinct forms of human thought which, in their appearances, have large autonomous areas of discourse and application that have very little to do with each other.

11.1 Saving the Appearances

Of course, two entities, A and B, need to be distinct in order to converge in the first place; I am not missing that logical point. In fact, I am going to play on it. For now, I am trying to draw attention to the methodological point that any interesting account of convergence ought to keep these opposing pulls firmly in view. Otherwise, it is all too easy to trivialize the issue. In our eagerness to locate convergences, we might be uncritically ignoring significant *non*-converging aspects of the disciplines. An account of convergence ought to leave enough degrees of freedom for the disciplines to diverge. Working with opposing intuitions thus severely constrains the scope of an interesting account. That is where the issue is

---

This is a thoroughly revised version of a paper published as Mukherji (2006).

© Springer Nature Singapore Pte Ltd 2017
N. Mukherji, *Reflections on Human Inquiry*,
DOI 10.1007/978-981-10-5364-1_11

most challenging. Let me illustrate this point by sketching some possible approaches to the issue that I reject, precisely because they do not obey the methodological constraint just suggested.

For example, it is possible to draw on wide and common notions of literature and philosophy such that a convergence could be seen in almost any instance of human discourse. With such wide notions, no reflective articulation of human thought can fail to display aspects of literature and philosophy wherever they are coming from. I myself have often been struck with the quality and articulation of thought when conversing, say, with a Santhal labourer working in my garden, or with a Bihari rickshaw-puller taking me to the market. One is impressed with the abstract nature of the opinion, subtle interpretation of common experience, rational character of the position defended, creative enumeration of choices at hand, uses of irony, metaphor, imagery, deliberate ambiguity, apt idioms, analogies, and the like.

It is difficult to conclude, from such ubiquitous presence of human grandeur (a phrase to which I return), that the *disciplines* of literature and philosophy have converged. These are not the sort of examples we are looking for when addressing the issue. One can see how the methodological constraint is already working to enable us to set these very general examples aside. The generality of these examples is exactly the trouble; since literature and philosophy converge *everywhere*, it is hard to see how to keep their appearances distinct.

A more subtle and involved application of the constraint arises as follows. One could approach the issue of convergence by listing some typical areas of convergence, and then attempt to give a unified account of what makes these areas possible. As a starter, one could suggest three kinds of items where the disciplines converge in appearance.

(A) Items of literature which have a distinct philosophical 'flavour', or have a philosophical point to make, or address a philosophical issue, *inter alia*. One could think of some of the novels and plays of Sartre, poetry of Sri Aurobindo, 'metaphysical' poetry of Donne, some later poetry of Tagore, some of the fiction by Milan Kundera, Kamal Kumar Majumdar, and the like.

(B) Works of philosophy which have a distinct literary 'flavour', which attempt to exhibit some literary style, solve some problems of narration, and so on. Plato's *Republic* is a popular example in this area. Bertrand Russell's work could be another. I am told that David Hume's work is sometimes taught in British schools as part of the English curriculum. There must also be similar examples in other traditions, for example, French or Italian. However, use of philosophical texts, for example, *Vedantin* and *Nyaya* texts, as part of standard programmes in Sanskrit literature won't be a good example. These are explicitly used as *Darsana* (that is, philosophical) texts, rather than *Sahitya* (that is, literature). It is an interesting example which needs further examination.

(C) Some items which typically blur the literature/philosophy distinction—sometimes by design. Two varieties immediately come to mind. One could cite the so-called 'Essays', say, by Charles Lamb or Amiel. A more voluminous variety could be the work of continental thinkers such as Heidegger, Lacan, Derrida, Barthes and the like. I will have a little more to say on this last variety in a moment.

I am not denying that each of these categories require detailed investigation. As we will see later, these categories could be viewed as more *explicit* examples of an underlying unity that binds literature and philosophy in a large variety of cases. The present point is that we should not jump to any conclusions about this underlying unity, if any, from these examples alone. Even if we take a *cumulative* enumeration of the items just listed, the enumeration falls hopelessly short of the vast bodies of literary and philosophical works. In other words, the enumeration excludes what might be taken to be (most of) the paradigmatic examples of the disciplines. For example, where do we place Aristotle, Nagarjuna, Descartes, Kant, Hegel, Frege, Wittgenstein, Carnap and Quine from the philosophy side, and Homer, Kalidasa, Shakespeare, Milton, Dante, Goethe, Tolstoy, Dostoevsky, Baudelaire, Jibanananda Das, D.H. Lawrence, Premchand and Manik Bandopadhyay from the literature side, in the suggested categories?

Again, I am not suggesting that we aim either for an all-inclusive account, or none. Given the complexity, variety and richness with which these ancient disciplines have unfolded, it is unrealistic to hope for an exhaustive account. Yet, for an account to have at least some semblance of empirical significance, we will expect it to cover a large spectrum of work with indications about how to conceptualize this spectrum. Towards the end, I will return to the issue of what empirical significance might mean for the project at hand. For now, we will at least expect a much wider coverage than what the items listed above promise. In short, following the methodological constraint, we can see that while the first group of examples—the gardener and the rickshaw-puller—makes the object of inquiry uninterestingly large, the second group consisting of items (A), (B) and (C) makes it embarrassingly small.

As a notable aside, notice that the preceding line of thinking allows us to attach only limited significance, for the issue at hand, to the body of work generally known as *continental thought*. One may doubt either the philosophical or the literary significance of this body of work precisely because much of this work blurs the disciplinary distinctions. One approaches literary and philosophical works with distinct states of mind, expectations, preparations, and the like. When one approaches, say, some of the work of Martin Heidegger with philosophical expectations, one finds some vaguely Aristotelian themes intertwined with socio-cultural observations. When one approaches them with literary expectations, one finds a rather heavy narrative style, laced with metaphors and deliberate ambiguities, which is reminiscent at once of Greek tragedy, classical German poetry and stream-of-consciousness literature.

One wonders, 'Why don't I read the untarnished originals, say, Aristotle or Sophocles or Goethe or Joyce themselves at *separate* moments of reflection and inquiry?' We wonder thus because we find *these* authors, just mentioned, to be firmly anchored in different reflective origins and discourses which, individually, offer insufficient illumination on Heidegger. In saying all this, I am not ignoring the individual significance of the Heideggerian *genre*. I am just questioning the significance of the genre for approaching the *project at hand*, for I suspect that if this genre is to be viewed as the central example of convergence, the issue will immediately lose much of its interested audience. The audience will lose interest because the issue of convergence is centred on something which is neither paradigmatic philosophy nor paradigmatic literature.

I am harping on this (what seems to me to be an) obvious point because I suspect that much of the current interest in the convergence issue, as a matter of popular fact, origins from the presence of this genre. There is no doubt that this range of discourse has allowed the more active and significant interaction between literature and philosophy: philosophers reflecting on literary works and authors, literature people reflecting on philosophers. Thus, whenever my literature friends feel interested in matters philosophical, they invariably have in mind something that emanates from Nietzsche, Heidegger, Derrida, Lacan or Foucault. The more curious of them may sometimes venture into later Wittgenstein, Thomas Kuhn or Hilary Putnam.

I do not know of any that would actually venture into the work of—not that they need to—early Wittgenstein, Rudolph Carnap, Alfred Tarski, Peter Strawson, Donald Davidson, Michael Dummett, Jerry Fodor or Saul Kripke. In fact, to push the point, it is often complained that these last named somehow fail to qualify as 'genuine' philosophers precisely because they fail to submit to expectations of convergence. Interestingly, a roughly similar distribution obtains for Indian philosophy as well. One reads the Vedas, the *Upanishads*, the *Gita*, perhaps bits of *Sankhya* and *Yogasutra*s, but not Dharmakirty, Vacaspati, Kumarila Bhatta, Gangesha or Raghunath. With such choices of convenience, the issue of convergence virtually disappears.

Several moderately surprising consequences seem to follow from these methodological considerations. First, it is unlikely that an interesting account of convergence could be reached by focusing on converging examples alone; thus, the challenge is to find convergence of literature and philosophy for those paradigmatic cases which appear to be *non*-converging, say, twentieth-century analytic philosophy and nineteenth-century Russian fiction, something like that.

Second, this conception of the project for *diverging visibles* suggests that convergence, if any, lies deeply buried in the collective invisibility of literature and philosophy; so perhaps it is *not* to be found in their textures, shapes and discourses, the properties that define their appearances. The search for convergence is thus a search in the subterranean. It could well be the case that the limited and dispersed examples of convergence cited above are in fact the more explicit manifestations of this underlying unity, just as natural geysers manifest deep and widespread convulsions in the bowels of the earth. In what follows, I will be concerned only with first-order examples of literature and philosophy. That is, I will not be concerned with disciplines such as literary criticism, philosophy of literature and aesthetic theory.

11.2 The Promising Point

One of the subterranean themes which I wish to examine in this chapter may be brought out with the help of a scene in Henry Miller's novel *Nexus*. In this scene, a hapless unpublished young writer walks into the dingy, foul-smelling office of an ageing lawyer. The writer is a destitute, and is forever looking for some money for his next meal. The lawyer is moderately successful in his profession, but is a failure in every other aspect of life. These two lost souls occasionally meet to share their grief and hopeless dreams. At some point during the dreary conversation that ensues, the lawyer, *not* the writer, remarks,

> Dostoievsky explored the field in advance, and he found the road blocked at every turn. He was a frontier man, in the profound sense of the word. He took up one position after another, at every dangerous, promising point, and he found that there was no issue for us, such as we are. (Miller 1966, p. 32, from one of the speeches by John Stymer)

The interest in these remarks for the project at hand is that Miller places them in the mouth of a common lawyer. It is not a professional remark of a literary critic, or of a philosopher drawing on Dostoievsky to make a metaphysical point. It is a remark of a sinking ordinary soul trying to find some grip on life in the company of another equally desperate human. And they both turn to Dostoievsky in their hour of suffering, because they think of Dostoievsky as a 'frontier man', in the 'profound sense'. Dostoievsky is credited with having 'explored the field in advance' to cover positions at 'dangerous, promising points'. By handing over the perplexities of their lives to Dostoievsky in this way, the writer and the lawyer find solace in the thought that 'there was no issue', after all. As if the dingy confines of the office in a drab block of apartments opened up to allow a larger, clearer vision of life.

Examples like these seem to justify the very common adage that literature provides a deep and sustainable understanding of the human condition in its totality and complexity. There is a widespread belief in Bengal that some words can always be gleaned from the poetry and songs of Tagore to lend clarity to any momentous point in life. It won't be surprising if a similar view is held of Sophocles, Dante, Shakespeare, Wordsworth and others in the West. Examples can be easily multiplied beyond 'official' literature to include fables, epics, *katha*s, folk songs and the like. Literature, then, is commonly seen as a guide, a source of practical wisdom, a master text of the human condition itself. This vintage role of literature as a source of *understanding* is the theme of this chapter.

The prevalence of an adage does not necessarily make it true. As societies and cultures change, there is a need to re-examine an adage of such general scope at every turn in history. In fact, certain contemporary developments might suggest that the adage is false, or at least that its significance is nearly obsolete. One recalls the classical opposition between religion and science, where it was held that religions were significant in enabling us to construct convenient world-views as long as insufficient scientific understanding of the cosmos prevailed. As 'enlightenment' dawned and science progressed, religions began losing their ground insofar as the *understanding* of the Universe is concerned. We no longer need to turn to religions to learn about the origin of planets, the beginning of life, and the growth of plants.

Similarly, it could be argued that we leaned on literature for ages due to our profound scientific ignorance of matters concerning humans. Once this state of ignorance begins to disappear, so does the relevance of literature in our common lives. Literature may continue to be a source of titillating entertainment, of technical virtuosity, a platform for display of verbal power; but its use as a source of understanding may not last as the human sciences, especially what is currently labeled *cognitive sciences*, grow.

This new opposition is well illustrated in the remarks of the philosopher Daniel Dennett, who is a leading philosopher wedded to current cognitive science. After explaining the truly remarkable phenomenon of virus replication, the exact chemistry of which is known by now, Dennett (1995, p. 203) concludes,

> An impersonal, unreflective, robotic, mindless little scrap of molecular machinery is the ultimate basis of all the agency, and hence meaning, and hence consciousness, in the universe.

Dennett believes that to think that the explanatory scope of 'molecular machinery' is limited only to the biophysical nature of humans is to believe in 'skyhooks' and in divine mysteries, rather than in the firmly grounded 'cranes' of science. In fact, the first part of the book, from which the preceding remark was cited, deals at length with the opposition between religious doctrines and the theory of evolution proposed by Darwin. Any perceived challenge then to the programme of understanding humans as scraps of molecular machinery invites the charge of obscurantism.

Given the growing postures of cognitive science, we have to ask afresh: What is the specific need of literature in human life; what does it do for us? As Jurij Lotman has wondered: why does every human community form literature? Moreover, we need to find out if the role we classically ascribe to literature could fall within the scope of cognitive science at all.

11.3 Limits of Science

At this point, I find it instructive to examine a remark made by Noam Chomsky some years ago. Chomsky suggested that we ought to turn to literature and the arts for deeper and comprehensive understanding of the human condition (Chomsky 1977). The content of Chomsky's remark, as we saw, is fairly commonplace in that we need not cite any authority to convince us of its value. However, the context and the authorship of the remark are of some interest for the theme at hand.

Noam Chomsky may be thought of as one of the principal architects of cognitive science. According to a very popular survey of the discipline about 30 years ago, cognitive science has, 'in a mere handful of years, discovered more about how human beings think than we had previously learned in all of our time on earth'

(Hunt 1982, p. 13). As cognitive science promises 'the systematic inquiry into our thinking selves', it establishes the fact that 'we are infinitely more intricate and remarkable than we have ever realized'. With these introductory remarks, the book surveys 'discoveries' not only regarding symbol-using ability of the mind, visual thinking, natural reasoning, metacognition and self-awareness, problem-solving, and the like, but cognitive styles, aesthetic judgements, metaphorical thinking, nature of creativity, and similar things as well (see also Chaps. 5 and 6, *this volume*). I am not concerned here with the merits and the actual results of cognitive science. Let us agree that cognitive science is, in fact, a major intellectual achievement unparalleled in the recent history of thought. The current interest is that much of the motivation for this discipline ensues from Chomsky's own work on language and mind.

Thus, Chomsky was asked, a little before the publication of Morton Hunt's book, whether in time cognitive science could possibly unravel the ultimate mystery of being a human. Chomsky replied as above, suggesting that a comprehensive understanding of the human condition could not be the agenda for science at all. In fact, according to Chomsky, that agenda is already classically addressed in the literature and the arts. When we focus on verbal articulation alone and, thus, set music, painting, sculpture and the other arts aside, it follows that literature is the *only* form of thought that pursues the agenda.

No wonder Daniel Dennett was duly upset about this apparent *volte-face* of a fellow cognitive scientist. He wrote, 'it is a great irony' that 'Noam Chomsky, automata theorist and Radio Engineer' 'was all along the champion of an attitude towards science that might seem to offer salvation to humanists'. 'Chomsky has argued', Dennett continued, 'that science has limits and, in particular, it stubs its toe on the mind.' So Dennett complains that, according to Chomsky, 'the only way would be the novelist's way—and he much preferred Jane Austin' to, say, Charles Darwin (Dennett 1995, pp. 386–7).

What explains Chomsky's 'revisionist' views in these matters? As far as I can see, the answer to this puzzle lies in Chomsky's views on the scope of science. 'Science', Chomsky says elsewhere, 'is a very strange activity' because

> It only works for simple problems. Even in the hard sciences, when you move beyond the simplest structures, it becomes very descriptive. By the time you get to big molecules, for example, you are mostly describing things. The idea that deep scientific analysis tells you something about problems of human beings and our lives, and our interrelations with one another and so on, is mostly pretence in my opinion—self-serving pretence, which is itself a technique of domination and exploitation. (Chomsky 2000, 2)

So the rough image is that physics basically ends with free electrons, chemistry with simple proteins, biology with a few dozen cells, and so on. The rest of the massive scientific-technological enterprise is a mixture of heuristics, common sense, tradition, local ingenuity and craftsmanship, as well as a fair amount of propaganda and hand-waving.

To see what could be at stake here, let us ask whether current research on cognition is able to tell a scientific story of behaviour at all, even if we set *human* behaviour aside. Consider, for example, the research on nematodes, a very simple organism with a few hundred neurons in all, meaning that people have been able to chart out their wiring diagrams and developmental patterns fairly accurately. Yet Chomsky reports that an entire research group at the Massachusetts Institute of Technology devoted to the study of 'the stupid little worm', some years ago, could not figure out why the 'worm does the things it does' (Chomsky 1994).

There are several lessons, some bordering on the political, to be learned here which I will not pursue. In particular, I am not questioning either the validity, or the supreme significance, of science. I am only anxious to draw attention to the curious fact that the deepest scientific understanding is inversely proportional to the complexity of the systems studied. It seems that, even if we set aside the understanding of 'true exercise of the creative imagination' which involves 'a mixture of madness', the study even of the 'lower form lies beyond the reach of theoretical understanding'. This is not necessarily an 'unhappy' outcome, Chomsky concludes. In that sense, science, like music, is best when it is essentially useless.

Before we proceed, let me note that, in sketching the preceding picture, I have taken *science* to mean essentially post-Newtonian modern science. One could object to this 'Eurocentric' conception of science, and decide to use the concept of science widely to characterize any reflective conception of the world (Nandy 1980). With this conception, it will be difficult to characterize any form of human thought to be non-scientific or prescientific. I, for one, fail to see what is achieved by this extended conception, except to miss the appearance of uselessness we just saw. The narrower conception of science seems to me to be at once harmless, empirically significant, and endowed with explanatory power as long as we do not attach primacy to the concept.

11.4   Tacit Mastery

Yet the profound fact is that, even though we cannot have scientific-theoretical understanding of anything beyond the simplest ones, we do have genuine understanding of *all* that matters, including that of 'human beings and our lives and our interrelations with one another', where 'scientific' understanding, as Chomsky told us, can only amount to self-serving pretence. In fact, we cannot fail to have such massive understanding if we are to function as humans at all. This is because of the obvious fact that humans are *cognitive agents*. Barring the so-called modular 'reflexes', almost every human act is an interpretive act, not just a response to stimuli. Hence understanding accompanies humans at every step. This ranges from the perceptions of a tree and a loved one to conceptions of future society and death. Experiments suggest that children, as young as six years, form fairly accurate conceptions of other minds, of things that are alive, and of systems that are 'natural' rather than artificial (Carey 1986; Keil 1989).

Beyond the domain of laboratories, I myself have often been struck with the quality and articulation of thought, as noted earlier, when conversing with simple common people in the street. Such ubiquitous presence of human grandeur

—'Cartesian common sense', as Chomsky calls it—clearly displays parts of the totality of the complex and largely successful understanding that humans routinely achieve. Is there some original connection then between our cognitive agency and the classical availability of literature and philosophy? Is it possible to construct a need for literature and philosophy beginning with the structure of cognitive agents?

Notice that the facts about human understanding noted above surprise us only when they are presented in objective, descriptive terms; that is, when they are brought to the fore, so to speak. In common life, we routinely behave with children assuming that they possess abstract knowledge of themselves, of others and of the world; in fact, we are worried when they don't. There is considerable current interest in the fact that autistic children seem to lack what may be called a *theory of mind* (Frith and Happe 1999); that is, these children cannot ascribe minds to others and, therefore, they fail to display much of cognitive behaviour that flows from such ascriptions.

The case is particularly interesting for the project in hand, since it is well known that over a millennium of philosophical discussions, and at least a century of scientific discussion, has led nowhere regarding a theoretical understanding of mind; yet normal children seem to accomplish the task effortlessly at a very early age. There is some evidence that children are not able to distinguish between definite and indefinite articles in their language until they have developed a theory of mind (McNamara 1984). Empirical evidence suggests that children routinely achieve a near-certain level of success by age three. This understanding, then, is largely manifest in our practical mastery of human conduct, not necessarily in a theoretical formulation of the conduct. This is one telling feature of the complex understanding embedded in common life which I am anxious to press here.

In formulating his conception of the philosopher, Strawson (1992, 7) suggests that

> the philosopher labours to produce a systematic account of the general conceptual structure of which our *daily practice* shows us to have a *tacit* and unconscious *mastery* (emphasis added).

I wish to focus on the notion of tacit mastery displayed in our daily practice. Philosophers, in this view, cannot even *begin* to produce a 'systematic account' unless the daily practices of common life have attained the vast mastery sketched above. Where does that mastery come from for philosophers to furnish an account of?

At this point, it is instructive to look at the structure of common life as it shows in our daily practices. Now even a glance at this vast system is beyond the scope of a single paper. Yet I wish to highlight some of the key features of the system to relate this part of the discussion to the issues at hand. The edifice is generally needed, of course, for us to be able to interpret and find our way through the overwhelming flow of experiences, feelings, desires, motivations, anxieties, and realizations—collective and individual. It seems that the structures so attained need to have at least the interrelated features of *connectedness*, *transcendence* and *locality* in order for us to be able to use the edifice in varying contexts.

Connectedness is required for us to form a coherent view of our immediate condition: conception of familiar objects, their relations with people's feelings including one's own, expectations and goals that form from there, the actions that flow with the resultant anxieties, frustrations and the ultimate broadening of horizons. *Transcendence* creates these horizons from the connected stock we already possess so that we can form conceptions of alternative choices, construct the past and the future, relate to other humans in terms of *their* choices and goals, so as to find a grip on the transitory and the ephemeral. *Locality* is required because we need to bring each of the unfolding perspectives to bear on the present and the finite space in which all our actions are necessarily located, namely, the domain of daily practices. Needless to say that these, and a lot more, are not *separable* aspects of our lives; we function as we do because these are in operation at once, and all the time. This rather breathless exercise, nevertheless, brings out two points.

First, the features of connectedness and transcendence together constitute an *open-ended* enterprise in terms of both reflections on the past and anticipations of the future. In other words, we could roughly identify this enterprise with *imagination* that helps us branch out beyond the present.[1] The crucial point here is that there is *an element of reflection* involved in the very exercise of imagination so conceived. The blanket term *imagination* as in, say, *she displayed great imagination*, signals two things: an object that is the product of imagination *and* an act that leads to the production of the object. To use contemporary jargon, we may view connectivity and transcendence as 'production systems' that, when implemented, generate the (novel) object; but implementation is an act that must be available to activate the system in the first place.

I am calling this implementing part *reflection* which must be a fundamental feature of the human condition for the exercise of imagination, that is, the ability to go beyond the present. So the 'mastery' we display in our daily practices cannot wholly be tacit in the sense of being unreflective. We saw earlier that this reflection, in general, cannot be of a scientific-theoretical kind, yet it fills every possible space of our daily practices. I am claiming that this ubiquitous feature of non-scientific reflection on the human condition is the essential source of all that is literary and philosophical in human thought. It is no wonder then that we find aspects of literature and philosophy wherever we look.

Nevertheless, the second point that emerges here is that the feature of locality, that is, the ability to harness the effects of imagination to our daily practices, severely restricts the scope of imagination. The constant pressure of daily practices

---

[1]The concept of imagination employed here has close connections with Karl Jaspers' notion of *transcendence* (Jaspers 1955), and Rabindranath Tagore's notion of *surplus* (Tagore 1931). In a more analytical vein, the concept is close to the contemporary notion of *possible worlds* (Stalnaker 1984). However, my contention that the logical theory of possible worlds is really a theory of imagination, is something that analytic philosophers will find hard to swallow. I can only point out that possible worlds are, in effect, networks of thoughts which are *constructed* from a *factual* basis. This applies to both the accessibility interpretation and the counterfactual interpretation in the technical literature on possible worlds.

—getting along with common life, so to speak—allows only so much room for imagination as needs to be harnessed for the task at hand. Let me stress that locality is *not* a limitation; it is in fact part of an efficient design that enables us to minimize reflective attention to maximize the efficiency of practice. The most efficient system, of course, is the *reflexive* one which involves no reflections at all. But then all actions become mandatory and no choices ever arise. Strawson is right then if we take the 'tacitness of mastery' to mean that, ordinarily, reflections do not dominate daily practice.[2] The fine edge we just supplied to support the notion of tacitness suggests that the interplay between reflections and tacitness is one of grades for cognitive agents like us. The image is that reflection and tacitness blend into each other with uneven effects throughout the system.

Let me gather some of the key points reached so far to lend some clarity to the issues raised. The vast and intricate web of common life, we saw, is not open to scientific-theoretical understanding. Yet, as cognitive agents, we cannot fail to reach some form of understanding of common life in its totality in order to lead it. In some sense, thus, the common life awaits our grasp of it. The reflective part of human imagination is the crucial ingredient for this achievement. However, given the hectic character of daily practices, reflection plays a limited locality-driven role in such practice. There is a paradox in this picture whose solution, in my view, comes from literary and philosophical traditions.

The paradox arises as follows. Given that reflection plays a limited role in our daily practices, and that reflection is the crucial element for grasping common life, it follows that common life is typically led without being fully grasped. We are beginning to confront a situation in which even tacit mastery must have severe limits if large parts of common life are beyond reflective grasp. But that would mean, in effect, that we lead much of our common lives *without* tacit mastery—by dint of 'instinct', so to speak.

Morals aside, the basic problem with any instinct-based picture is that it simply does not square with the idea of cognitive agency. Centuries of philosophical and psychological speculations could not untie the knot that gets formed once we foster even a limited instinct-based story to explain our daily practices. There is something to be learned from the fact that even human sexual practice, to mention just one of them, is an elaborate reflective enterprise. This is not to deny that there are reflexes elsewhere in the system; yawning when feeling sleepy isn't a cognitively-geared daily practice in the Strawsonian sense under discussion here. In short, it is simply incoherent that we are at once cognitive agents in some respects and not cognitive agents in other respects, wherever we draw the line across our common lives. So the paradox is that although, by the nature of daily practices, we are pushed towards an instinct-based picture, the picture cannot be true even in part *insofar as daily practices are concerned*.

The tension between imagination and locality, sketched above, thus assumes an alarming character, and it puts great strain on the very notion of cognitive agency.

---

[2]See Fodor (1983) and Pylyshyn (1984) on *encapsulated* systems and *central* systems.

The open-enedness of human imagination abruptly faces closure in view of the absence of reflection in much of cognitive life. I hope it is clear that the problem is not so much the absence of reflection, but the lack of access to a comprehensive understanding of common life. The absence of reflection seemingly denies the only route to that understanding. How then do we retain the full-blooded notion of cognitive agency, and avoid an instinct-based picture of our daily practices?

11.5 Division of Understanding

The only solution I can think of is that we must have access to reflective resources *other than* what we individually exhaust in conducting our common lives. The thin edge on which these resources must be placed requires that two opposing conditions are simultaneously satisfied: given that the daily practices are our only cognitive arena, so to speak, these resources must arise from daily practices themselves. Yet, given the general absence of non-local reflection in ordinary daily practices, they cannot arise from our *ubiquitous* daily practices. The only way this can happen is that there be *non*-ubiquitous daily practices geared to a comprehensive understanding of common life itself.

What I have in mind is a sweeping generalization of a very special case once studied by Hilary Putnam (1975). Putnam's problem was that we routinely use different terms like *elm* and *beech* without being able to tell the difference between elms and beeches. The trouble is that our use (=daily practices) suggests that we know that *elm* and *beech* differ in meaning, yet we cannot tell what the difference is. Putnam's solution is that although we commonly do not know the difference, we know who knows the difference, namely, the botanist, to whom we can turn when required. There is thus a 'social division of linguistic labour' that distributes knowledge of language across layers of language-users. I am not suggesting that Putnam's insight works for theory of meaning (see Fodor 1994). My hope is that some version of it works analogically for the project at hand.

Notice that Putnam's solution requires that it is a part of the botanist's daily practice that he has a comprehensive understanding of flora. The availability of that understanding in the botanist's shelves is assurance enough for *us* to carry on with our diffused uses of *elm* and *beech*. Similarly, I am suggesting that the paradox of common life sketched above is solved because there is a social division of reflective understanding of common life. The 'experts' whose business it is to come to terms with the complexity of common life as part of *their* daily practices are the authors of literature and philosophy. The presence of literature and philosophy thus solve a fundamental dilemma for cognitive agents.

Notice as well that Putnam's solution requires that we make reference to the *cumulative practice* of botany when we refer to a given botanist. The reference, thus, is to the *discipline* of botany which a botanist represents by virtue of belonging to the tradition of botany. This enables a cumulative record of expertise to be available for common reference. What enables a botanist to belong to his discipline includes acquisition of specialized skills, rearrangement of daily practices such as spending more time in the library or the laboratory, access to the existing body of knowledge that one gains when those skills are mastered, and, most importantly, an ability to create new knowledge in continuity with the tradition. The presence of a disciplinary tradition, then, is an inevitable consequence of specialized reflective practices. Insofar as bodies of literature and philosophy represent the cumulative record of the specialized practice of coming to terms with the totality of common life, they are disciplinary traditions as well with all the features of a tradition just noted.

However, *this* specialized practice is markedly different from the practice of botany in at least one significant dimension. Kuhnian stories of science aside (Kuhn 1962), the continuity of the tradition of botany is a (more or less) linear one in that one stage of the tradition gets absorbed in the next without a trace. This is because the domain of botany can be studied in isolation from the rest of our practices; the botanist views flora as if 'from nowhere,' to use a phrase popularized by Thomas Nagel (1986).

The traditions of literature and philosophy that concern us here, however, have the entire common life as the domain, the exploration of 'one position after another, at every dangerous, promising point'. Further, the entirety of common life has to be grasped somehow from within one or other of the daily practices that locate the author. So every attempt at the total grasp of common life is from a point of view; in principle, there cannot be a view from nowhere. Thus, as the locations of authors change significantly, the points of view change accordingly generating, in turn, alternative conceptions of human life.

It seems to me that these requirements lead to a fairly accurate notion of a literary or a philosophical text in the standard sense. A body of work becomes a *textbook* in botany precisely because botany does not have alternative standpoints in the sense outlined. The discipline of botany does not have texts, although it typically has textbooks. Literary traditions do not have textbooks; they just have texts. This asymmetry between texts and textbooks seems to me to be a pleasing result since it is empirically significant. We have been able to extract an explanation, from more or less first principles, of the puzzling fact that the disciplines of literature and philosophy are critically marked by texts. In time, the presence of textual traditions outline entire cultures that seep into common life at various points to invigorate and change the character of daily practices themselves.

This picture of literature and philosophy with textuality at its centre requires closer scrutiny with respect to other forms of thought. I have no space here for the task. But notice, before we begin to suggest counter-examples to the idea, that if we find the relevant sense of textuality in, say, sociology, we have to make sure that: (i) the suggested textuality is not due to the literary and philosophical aspects of the discipline; and (ii) textuality is a *necessary* feature of the discipline, that is, the discipline retains its textual character even in its advanced forms, preventing thus the appearance of textbooks. Moreover, it is quite possible that some disciplines, say, music or the other arts, have a 'textual' culture because of the skill-part of the discipline, namely, you copy or practice traditional forms. It is not clear to me that there are different points of view in music or in painting that have to do with differing *conceptions* of common life (*this volume*, Chap. 10). Once we apply these tests, it seems to me that only literature and philosophy pass them.

I have argued that textuality, geared to a reflective but non-scientific under-standing of common life, is the more explicit form of the diffused understanding that guides common life in any case. Since the common life of cognitive agents is itself marked by sustained reflection, albeit constrained by locality, literature and philosophy, not unlike the radical activist, is firmly entrenched in common life; thus, nourishing it, nudging it, and getting enriched in turn. The marking out of this textual territory has been an exercise in the subterranean, as expected at the outset.

Nevertheless, the picture leaves enough degrees of freedom in the choice of daily practices, in the selection of focus and concern, and in the skills that accompany them, to allow marked differences in how literature and philosophy appear on the surface. I will briefly describe just one dimension along which literature and phi-losophy classically differ, for, I believe that most of the other dimensions—for example, literary works are works of art, philosophical works typically are not—can be traced to this fundamental dimension. As far as I can see, there are exactly two ways of coming to a comprehensive understanding of common life.

Recall that Strawson viewed philosophers as engaged in 'producing a general systematic account of the conceptual structures' displayed in our daily practices. Thus, philosophy is concerned with the conceptual basis of common life. This inquiry is essentially abstract in that it can only be conducted at a certain remove from common experience. Hence it lacks the colour and the tone of felt experiences that animate the common life. In that sense, the comprehensiveness of a philo-sophical account of common life is more like an aerial view of common life—mistakenly thought of as a view from nowhere. That is one way to understand common life.

Another could be to produce a more detailed account of the emotional and existential structures displayed in our daily practices. In this form, we confront the complexity of common life head-on, as it were. We start at any point and keep digging through as much ground as we can cover until we reach a coherent per-spective, narrating the view that unfolds from that angle. As long as these activities differ, the character of the texts will differ as well.

References

Carey, S. 1986. *Conceptual Change in Childhood*. Cambridge: MIT Press.
Chomsky, N. 1977. *Language and Responsibility: Conversations with Matsu Rona*. New York: Pantheon.
Chomsky, N. 1994. *Language and Thought*. London: Moyer Bell.
Chomsky, N. 2000. *The Architecture of Language*. New Delhi: Oxford University Press.
Dennett, D.C. 1995. *Darwin's Dangerous Idea*. London: Penguin.
Fodor, J. 1983. *The Modularity of Mind*. Cambridge: MIT Press.
Fodor, J. 1994. *The Elm and the Expert*. Cambridge: MIT Press.

Frith, U., and F. Happe. 1999. Theory of mind and self-consciousness: What is it like to be autistic? *Mind and Language*, 14(1): 1–23.

Hunt, M. 1982. *The Universe Within: A New Science Explores the Human Mind*. New York: Simon and Schuster.

Jaspers, K. 1955. *Reason and Existenz*. Translated from German by William Earle. New York.

Keil, F. 1989. *Concepts, Kinds and Cognitive Development*. Cambridge: MIT Press.

Kuhn, T. S. 1962. *The Structure of Scientific Revolutions*. Chicago, IL: University of Chicago Press.

McNamara, J. 1984. *Names for Things: A Study of Human Learning*. Cambridge: MIT Press.

Miller, H. 1966. *Nexus*. London: Panther.

Mukherji, N. 2006. Textuality and common life. In *Literature and Philosophy*, S. Chaudhury (Ed.). Papyrus: Kolkata.

Nandy, A. 1980. *Alternative Sciences*. New Delhi: Oxford University Press.

Nagel, T. 1986. *The View From Nowhere*. Oxford: Oxford University Press.

Putnam, H. 1975. The meaning of 'meaning'. In *Language, Mind and Knowledge, Minnesota Studies in the Philosophy of Science*, vol. 7, K. Gunderson (Ed.). Minneapolis: University of Michigan Press.

Pylyshyn, Z. 1984. *Computation and Cognition*. Cambridge: MIT Press.

Stalnaker, R. 1984. *Inquiry*. Cambridge: MIT Press.

Strawson, P. 1992. *Analysis and Metaphysics*. Oxford: Oxford University Press.

Tagore, R. 1931. *The Religion of Man*. London: George Allen and Unwin.

Chapter 12
Education for the Species

> *In this possibly-terminal phase of human existence, democracy and freedom are more than values to be treasured, they may well be essential to survival.*
>
> Noam Chomsky

Noam Chomsky's grimly titled book *Hegemony or Survival* (2003) opens with some observations of contemporary biologist Ernst Mayr, who is sometimes referred to as 'the biological giant of the twentieth century' (Foreman 2004, p. 24). After proposing a very reasonable notion of a species (de Queiroz 2005), Mayr (2001) held that about 50 billion species have appeared on this planet since the origin of life. He estimated that 'the average life expectancy of a species is about 1,00,000 years' (cited in Chomsky 2003, p. 1). Exactly one of these 50 billion species 'achieved the kind of intelligence needed to establish a civilization,' Mayr notes (cited in Chomsky 2003, p. 1). The civilization-forming intelligence of this species is the topic for this chapter.

From studies on sudden expansion of brain size (Striedter 2004), restructuring of the brain for emergence of language (Crow 2010), and proliferation of tools and other signs of culture, it is now estimated that the modern human species emerged roughly about 1,00,000 years ago (Tattersall 2012). Following Mayr's statistical rule, then, the species is possibly nearing its end. If the species becomes extinct due to the natural course of events, then there is nothing much we can do. But if the projected extinction is linked to its civilization-forming intelligence then a very different set of issues arise.

This is a revised version of a paper published as Mukherji (2016).

© Springer Nature Singapore Pte Ltd 2017
N. Mukherji, *Reflections on Human Inquiry*,
DOI 10.1007/978-981-10-5364-1_12

12.1   Sixth 'Intelligent' Extinction

We may hope to defy Mayr's doomsday scenario under the impression that the human species, apparently, has remarkable control over its destiny, precisely due to the 'kind of intelligence' with which it is endowed. Humans may feel reassured that this kind of intelligence will ultimately devise ways, technological and otherwise, to protect the species beyond its statistical limit. Unfortunately, the hope seems to lack foundations. Mayr's controversial estimate is not the only clue for his doomsday scenario. He proposed another perspective in which the prospect of premature extinction is in fact enhanced by the human kind of intelligence. It is just that the two scenarios seem to converge on the time left for the species.

Biologists suggest that there are two evolutionary scenarios that lead to the extinction of species. The first form of species extinction is called *background extinction*. This form of extinction happens due to background factors, such as low density of population, limited dispersal ability, inbreeding, successional loss of habitat, climate change, competition, predation, disease, and the like (Soulè 1996). There is considerable dispute about the life of a species undergoing inevitable background extinction. As noted, Mayr thought that species-life is as low as 100,000 years. Others calculate it between 1 million and 5–10 million years.

Biologists also list a second form of extinction—*mass extinction*—in which more than 50% of all species on earth, at a given point in time, are wiped out simultaneously due to some massive catastrophe. Biologists identify five events in the last half a billion years when such grand-scale extinction happened. The last of these—*the Cretaceous*—occurred when, 65 million years ago, dinosaurs and many mollusks became extinct, most probably due to the strike of a giant asteroid.

In either case, species become extinct due to what may be viewed as natural reasons that are external to the species. These occur in nature periodically due to circumstances beyond the control of the members of the species. In these cases of natural extinction on a geological scale, nothing much can be done in the long run, even if a variety of 'intelligence' and other favourable factors postpone the inevitable in the short run. At the current stage of knowledge, there is no definite prediction that the human species is about to become extinct due to the convergence of natural background factors or some catastrophic event, such as the striking of a giant asteroid.

The prediction, rather, is that, after a lapse of 65 million years, the conditions for another—sixth—mass extinction are maturing rapidly. The human species is most likely to disappear due to phenomena such as nuclear holocaust, massive environmental destruction, global conflict, including biological warfare, astronomical poverty, irreversible damage to food chains, and maybe even just the unavailability of potable water. The extinction of the species will most likely be caused by the suicidal behaviour of the species itself. As Chomsky puts it, *we* are the asteroid.

The author of *The Sixth Extinction: An Unnatural History*, Elizabeth Kolbert (2014) suggests in an interview (Drake 2015) that the factor of environmental degradation due to human recklessness alone has enhanced the rate of species extinction by more than 100 times the normal rate in just the last few hundred years. This is because, Kolbert argues,

We loaded the extinction rate with widespread hunting, we brought in invasive species. We are now changing the climate, very, very rapidly, by geological standards. We are changing the chemistry of all the oceans. We are changing the surface of the planet. We cut down forests, we plant mono-culture agriculture, which is not good for a lot of species. We're overfishing (cited in Drake 2015).

The list goes on and on. To emphasize, Kolbert's picture only includes extinction of other species triggering mass extinction. To this picture, we need to add factors like nuclear holocaust, global war, dislocation of food chains, massive famines, depletion of potable water, and the like, which more directly relate to the extinction of the human species itself.

Significantly, each of these doomsday scenarios is critically linked to the species' unique endowment of the 'kind of intelligence needed to establish a civilization' (Chomsky 2003, p. 1). No other species remotely has the ability to change the chemistry of the planet, and pollute much of the potable water on earth, by its own diligent effort in just a few hundred years, not to mention the ability to construct weapons of mass destruction, to which we will return.

As Mayr pointed out, there is no evidence that nature prefers intelligence over stupidity: beetles and bacteria, for example, are vastly more successful than the great apes, not to mention humans, in terms of survival. Looking at humans through this long lens of evolution, it could well be, Chomsky holds, that humans were a kind of 'biological' error, using their allotted 1,00,000 years to destroy themselves and much else in the process with 'cold and calculated savagery' (2003, p. 2).

The centrality of the notions of intelligence and stupidity brings the topic of the imminent extinction of the species within the broad domain of education. Hence, the title of this chapter.

12.2 Ideology and Hegemony

For his book *Hegemony or Survival*, Chomsky used the subtitle *America's Quest for Global Dominance*, suggesting that the prospect of human survival depends primarily on how humanity responds to the hegemony of the United States (USA). No doubt, with its absolute military control over the planet and the space around it, and its nuclear hardware capable of vaporizing much of the planetary system, the USA has represented the peak of the 'cold and calculated savagery' with which humans have proceeded to destroy themselves. Moreover, through the use of its military control, the USA has thwarted almost every effort to get the planet on some track of recovery. For example, in the last few decades, it has not only ignored the Geneva Convention on warfare and the United Nations (UN) resolutions on terrorism, but has also walked out of the Kyoto Protocol on the environment, the Anti-Ballistic Missile (ABM) treaty, and the convention on biological warfare, among others. There is some basis, then, for viewing US hegemony as a principal agent for the imminent extinction of the species.

However, the USA has not been alone in taking such actions. The ideology that governs US hegemony over the planet had precedents throughout the history of the Western world. As Chomsky (2005, p. x) notes, the German philosopher Martin Heidegger, rated by many scholars to be one of the greatest thinkers of the twentieth century, viewed Nazi Germany as the most 'metaphysical of nations'. After constructing the spectre of the Jewish–Bolshevik conspiracy to take over the world, eminent Western intellectuals thought that 'extreme measures' were necessary for 'self-defence'. 'As the Nazi storm clouds settled over the country in 1935', Chomsky continued, 'Martin Heidegger depicted Germany as the "most endangered" nation in the world, gripped in the "great pincers" of an onslaught against civilization itself, led in its crudest form by Russia and America' (Chomsky 2005, p. x). According to Heidegger, Germany stood 'in the center of the Western world', and must protect the great heritage of classical Greece from 'annihilation', relying on the 'new spiritual energies unfolding historically from out of the center'. Hence, the catastrophic war was needed to protect the 'great heritage of classical Greece' (Chomsky 2005, p. x).

When it was attacked by the Japanese in Pearl Harbor, the USA unleashed its own 'legitimate exercise of self-defense against a vicious enemy' (Chomsky 2005, p. xi) with a 1000-plane daylight raid on defenceless Japanese cities, culminating in the nuclear bombing of Hiroshima and Nagasaki. Chomsky notes:

> The paroxysm of slaughter and annihilation did not end with the use of weapons that may very well bring the species to a bitter end. We should also not forget that these species-terminating weapons were created by the most brilliant, humane, and highly educated figures of modern civilization, working in isolation, and so entranced by the beauty of the work in which they were engaged that they apparently paid little attention to the consequences. (2005, p. x)

As Chomsky has pointed out, the basic problem is much deeper and historical in character than the immediacy of a rogue state (Gettys 2014). Thus, even if the current neoliberal phase represents the 'extreme end of the traditional US policy spectrum', these policies have 'many precursors, both in US history and among earlier aspirants to global power'. 'More ominously', Chomsky continued, 'their decisions may not be irrational within the framework of prevailing ideology and the institutions that embody it' (2003, p. 4). This is the crucial point—these are rational decisions taken in a civilization mode, these are products of the most sophisticated thinking pursued for hundreds of years in great centres of learning. In that sense, there is a direct correlation between the culture of enlightenment and the untimely extinction of the species.

Beyond US hegemony, there is now growing concern that humanity might well be led to a species-terminating global war originating in West Asia. After bitter plunder and strategic warfare conducted by the West for over five decades, this region has now spawned powers, such as the Islamic State of Iraq and Syria (ISIS), that not only have the armed resources to acquire local state power, but have the determination to achieve global dominance just like Nazi Germany. In fact, their ideologies go beyond that of Nazism to actually seek the end of the world. Graeme Wood (2015) reports that

> we can gather that [ISIS] rejects peace as a matter of principle; that it hungers for genocide; that its religious views make it constitutionally incapable of certain types of change, even if that change might ensure its survival; and that it considers itself a harbinger of—and headline player in—the imminent end of the world.

It is instructive to note in this connection that the prospect of species termination is not restricted to avowedly hegemonic violent states and their ideologies. Thus, Chomsky mentions the apparently benign and peace-loving country, Canada, to understand the real scope of the concerned ideology. Speaking on the energy policies of the Canadian government under former Prime Minister Stephen Harper, Chomsky observed that

> It means taking every drop of hydrocarbon out of the ground, whether it's shale gas in New Brunswick or tar sands in Alberta and trying to destroy the environment as fast as possible, with barely a question raised about what the world will look like as a result. (cited in Lukacs 2013)

Needless to say, such destruction of the environment continues across the world, including in India. And the destruction of the environment puts immense pressure on available resources such that access to the remaining resources enhances the prospect of catastrophic war.

12.3 Indigenous Resistance

Most importantly, for our purposes, Chomsky also sketched an alternative to these entrenched ideologies by applauding the resistance against these policies raised by the indigenous people congregating at the margins of Canada's much-flaunted multicultural society. 'It is pretty ironic', Chomsky remarked, 'that the so-called "least advanced" people are the ones taking the lead in trying to protect all of us, while the richest and most powerful among us are the ones who are trying to drive the society to destruction' (cited in Lukacs 2013).

The general lesson is hard to miss. Notice the expression 'all of us'. The resistance by the indigenous people to the extraction of hydrocarbons not only saves the environmental niche of these people in New Brunswick and Alberta, but is in fact protecting all of us, the species. In contrast, the rational choices enforced by the ideologies and the institutions controlled by the rich and the powerful are driving the human race towards extinction. It is therefore an issue about the salient authorship of knowledge.

The issue of knowledge emerged vividly nearer home in the jungles surrounding the Niyamgiri Hills in the state of Odisha. These hills contain about 1.8 billion tonnes of high-grade bauxite, the source for aluminum, which a mining giant—euphemistically called *Vedanta*—wants to extract to feed into giant factories built on this land. As they were pushed out of the plains by the thrust of mainstream civilization, the local poor, mostly tribals, had lived on this hilly land for thousands of years. After years of resistance by them—and much manipulation and show of muscle by the state, financed by the mining oligarchy—the government was compelled to organize a referendum for 12 carefully selected villages when the fate of hundreds of villages was involved (Kothari 2015; Vanaja 2014).

As one of many moving studies reports (Bera 2013), using the democratic and peaceful resource of their own *panchayats*—units of local self-government—village after village gathered *en masse* amid heavy security cover of central paramilitary and state forces. Ignoring the guns and bayonets, 'unlettered' forest dwellers—Dongria Kondh and Kutia Kondh tribals, and Gouda and Harijan non-tribals—spoke of a religion embedded in the hill's pristine ecology. They told the district judge, appointed observer to the meetings by the apex court, that mining will destroy their god and their source of sustenance:

> Over 100 perennial streams, fruit trees such as jackfruit and mangoes, spices like turmeric and ginger, wild roots, tubers and mushroom; apart from the land for shift and burn cultivation—*dongar*—where they grow an enviable mix of native millets, pulses and oil seeds (Bera 2013).

Having said this, each village unanimously rejected the Vedanta project. The Niyamgiri Hills survived. For now, mark the word *unlettered*, as was used by the reporter. The people themselves ratified this perspective of illiteracy. Tunguru Majhi, a Kutia Kondh tribal, declared at the Kunakadu *palli* village council meeting,

> We will die like Birsa Munda and Rindo Majhi [both Munda and Majhi led tribal uprisings against the British] if you don't give up now. We are a *murkhya jati* [illiterate people] who will never listen to you (Bera 2013).

This illiteracy, the absence of letters, the stupidity of the ancient belief in a caring and protecting god of the hills, might just provide the answer to the question of whether or not the species will survive after all.

12.4 Questioning Liberal Pedagogy

Recall that when he mentioned the resistance by the indigenous people of Canada, Chomsky used the expression 'so-called "least advanced" people' (Lukacs 2013). He is not only referring to their action of resistance, but pointing at their intellectual achievement, without which the action of resistance would not have followed. In contrast, the 'rational decisions' reached by formidable intellectuals serving the rich and the powerful lead the species to the verge of extinction. The contest is, therefore, between two opposing systems of knowledge in two different *intellectual* traditions.

Moreover, Chomsky's contrast between the two traditions implies that, in a crucial historical sense, elite intellectual traditions have failed the species, while the indigenous traditions, in almost total isolation from the elites, open the opportunity for the continued survival of the species. In the same historical sense, then, survival of the species now depends on incorporating marginalized indigenous systems of knowledge into the mainstream. At the same time, there is a need to severely critique and progressively replace entrenched aspects of elite intellectual traditions, which have ruled the world for at least the last few hundred years in the garb of liberal pedagogy.

What does this scenario mean for education policy? What does it mean exactly to prioritize and adopt the knowledge-systems of the *murkhya* to save the species and the planet? In the limited space available to me here, I will focus on the prospect of incorporating indigenous knowledge in the mainstream education policy. In the process, I will be barely able to touch upon the related, but wider issue of dispensing with much of the current liberal curriculum that generates the mindset for plundering the planet.

Ever since liberal education became the agenda at the turn of the last century, education of the poor and the marginalized has concerned a range of progressive thinkers. I will briefly touch upon two of them—Rabindranath Tagore and Paulo Freire—to suggest why these responses to the issue of the survival of the species are inadequate. There are two reasons why I wish to focus on these authors. First, given the historical problems of modernity, there is already growing awareness that Western liberal education has not lived up to its promise of enlightenment, as noted above. In that context, it is of much interest that both Tagore and Freire are non-Western critics of Western elitism and are well-known for their views on education policy. Second, both direct their attention to the education of the marginalized as a form of universal welfare. How do their apparently egalitarian liberal views fare with respect to the issue of indigenous knowledge?

12.4.1 Education for Fullness

Tagore was deeply troubled by the extreme elitism of the British-enforced education system that catered only to the children of the privileged. As is well known, he was also deeply critical of the kind of education that was imparted, the rote learning that Freire later identified as the 'banking' method. Instead, Tagore advocated an enlightened and elaborate version of education for fullness: *sarbangin shiksha*. This included not just the education of the intellect, combining the most universal aspects of Western and Eastern high culture, but also the education of feeling for the other that extended to feeling for nature and cosmos. In this sense, he criticized the one-sidedness of an education that only imparted bookish knowledge in a narrow, pragmatic sense. His conception of education did not reject the ideals of Western enlightenment, but sought to embed it in a wider conception of learning that, he thought, embraced the whole human (Mukherjee 2013).

There is no convincing evidence that the knowledge-systems for 'fullness' that constituted Tagore's conception of *sarbangin shiksha* included the knowledge-systems of the unlettered even in its margins. So, his lament about the absence of the poor from the field of education may be viewed as a 'humanitarian' lament, not really a 'humanistic' one, to use a distinction suggested by Freire and to which I return.

In fact, there is evidence that Tagore viewed the poor and the marginalized as ignorant, dull and voiceless, to whom language needs to be imparted, and hope needs to be aroused in those broken hearts.[1] And the knowledge that is supposed to enlighten the poor is the high culture knowledge already imparted to the elite. Needless to say, this task of pulling the poor out of their misery through *sarbangin shiksha* required novel educational practices such as teaching in the mother tongue, using local flora and fauna as examples, active agency of the learner, the *tapovana* model of shunning bounded classrooms and holding learning sessions in the open air, etc. Yet the knowledge that was so imparted consisted of the products of the elite high culture, from the Upanishads to modern science, via literature, art and sophisticated musical forms.

I think the point about the ultimately elitist character of Tagore's otherwise enlightened conception of education can be strengthened with an example of novel educational practice followed in Tagore's school. I could not locate any official document for this, but I can recount this curious practice from my own experience as a student in Tagore's school at Santiniketan. Every afternoon, children from Patha Bhavana were transported in the university bus in batches to Silpa Sadana at the rural setting of Sriniketan, the location for rural education and reconstruction. There, we sat down on the floor to learn about woodcraft, papier mâché, basket weaving, lac work, etc., from the ill-clad and impoverished, but highly skilled village artisans. During that period of active hands-on learning, some of the rural folk were our teachers. Our education, thus, included some of the knowledge-systems of the unlettered, and a reversal of class roles. No wonder this novel education practice was soon abandoned due to logistical reasons.

Yet the point remains that the appreciation and adoption of rural culture was restricted to the 'crafts' of a folk nature. Elite, high culture still formed the central ingredient for the development of sensitive intellect. Similarly, farmers are sometimes consulted about various agricultural practices such as variety of seeds, condition of soil, multiple cropping, organic fertilizers, etc. This is the traditional domain of the unlettered where knowledge is accumulated through sheer practice over centuries. Beyond this, rural culture (not to mention tribal culture)—except 'folk art'—is not ascribed any enlightenment value. The tribals, the indigenous people, are not even in view. They are curiosities hiding in hills and forests.

---

[1]Tagore, in the poem titled *Ēbāra phirā'ō mōrē* from the volume *Chitra*:
Ē'i-saba mūr.ha mlāna mūka mukhē ditē habē bhās.ā
ē'i-saba śrānta śus.ka bhagna bukē dhban.yā tulitē habē āśā.

12.4.2 Humanistic Education

Several decades later, Paulo Freire, in his classic work, *Pedagogy of the Oppressed* (1970/2005), addressed the issue of resistance to the ideologies and institutions of the elite more directly. The task for education, he felt, was to reverse the process of dehumanization in which the oppressed found themselves:

> The struggle for humanisation, for the emancipation of labor, for the overcoming of alienation, for the affirmation of men and women as persons... is possible only because dehumanisation although a concrete historical fact, is not a given destiny but the result of an unjust order that engenders violence in the oppressors, which in turn dehumanizes the oppressed (Freire 2005, p. 44).

Following George Lukacs, Freire elaborates that a revolutionary educational practice aims to 'explain to the masses their own action', to clarify and illuminate that action, both regarding its relationship to the objective facts by which it was prompted, and regarding its purposes (53). The more the people unveil this challenging reality, which is to be the object of their transforming action, the more critically they enter that reality. In this way they are 'consciously activating the subsequent development of their experiences' (53). Freire insists that this form of education is essentially pre-revolutionary, such that the oppressed can proceed to a revolutionary overthrow of the unjust order. Freire, thus, goes beyond Tagore to view education not only as a humanitarian mode to include the oppressed, but as one which triggers humanization of the oppressed by enabling them to erect the other side of the barricade. Let us call this mode of education the *proletarian* mode.

It is unclear if the envisaged overthrow of the unjust order will in fact enhance the prospects for the species as a whole. The humanized education achieved through the struggle of the working masses will no doubt usher in an era of proletarian freedom. But, will it ensure survival for all? The answer will depend on the content of the proletarian mode, the knowledge-systems so advocated. Here, the prospects do not appear to be as revolutionary as the emancipation of a section of people.

There is little evidence that pre-revolutionary education practices among the masses, undertaken by revolutionary forces, address the issue raised here. In his writings, Freire makes frequent references to politico-educational work of Mao during the pre-revolutionary phase. Following these examples and their implementation during, say, the struggles in Yan'an and Vietnam, certain forms of educational practices have emerged. For example, following lessons from Vietnam, Maoists in India have organized Young Communist Mobile Schools (or Basic Communist Training Schools), which host select groups of 25–30 tribal children in the age group of 12–15 years.

These children receive intensive training for six months in a curriculum consisting of basic concepts of Marxism–Leninism–Maoism, Hindi and English, mathematics, social science, different types of weapons, computers, etc. (recall their age group). Needless to say, lessons are conducted in Gondi, and local song and dance forms are used to motivate the children. Beyond this, there is no evidence that the ancient knowledge-systems of the tribals form any significant part of the curriculum, even though the pupils concerned consist entirely of tribal children. In fact, much of the curriculum, including lessons in modern science go directly against the foundations of tribal culture; especially, weapons training involving not bows and arrows, but automatic rifles, light machine guns, high-powered explosive devices, and the like (Mukherji 2012). While the children in mainstream India sit through modernist curricula under the aegis of not-so-subtle capitalist propaganda, tribal children sit through roughly the same curriculum, even if they have been asked to wear Maoist lenses. Education is imparted in the proletarian mode, not in the indigenous mode. It is difficult to dispel the impression that modernist educational thinking has deeply penetrated even the most revolutionary minds.

12.5   Concluding Remarks

It seems plausible to hold, then, that the most progressive, enlightened forms of thinking on education fail to offer a sustainable perspective on the survival of the species. In some grim historical sense, the prospects seem irreversible because the so-called enlightened conception of knowledge, which is primarily responsible for bringing the species to the brink of extinction, is uncritically assumed to be the only one we have. In fact, liberal education, with its species-terminating edifice of knowledge, is often ascribed absolute value, since any alternative form of education is viewed as either inconceivable or politically incorrect.

What is missed in these universalist proclamations in favour of liberal education is that an entire range of indigenous knowledge-systems have existed simultaneously, but in almost total isolation from the modernist liberal knowledge-systems. These are not 'primitive' or 'infantile' systems of knowledge requiring further stages of development. These systems are current 'adult' systems of knowledge with their own high culture that have been sustained in favourable environmental niches for thousands of years. If liberal education can claim its historical validity by referring back to the *Vedas*, *Sutras*, Euclid and Plato, so do the indigenous systems, except that their classical heritage has remained unnamed in the absence of global propaganda. These systems define the alternative forms of what it is to be human as a species. The only problem is that these systems, with their construction of God of Niyamgiri and reverence for rivers, are viewed as inconsistent with the modernist outlook. But that certainly is a problem for the modernist, not the Dongria Kondhs.

In other words, a real solution to the issue of survival requires that humans learn to progressively forget—or, at least, engage in severe criticism of—the knowledge-systems currently advanced in the most dominating centres of learning. If indigenous knowledge-systems, currently resisting extraction of hydrocarbons and bauxite from forests, are our primary route for survival, every bit of knowledge beyond indigenous knowledge must be subjected to serious critique for their relevance.

I am aware of the possible inconsistency in what I am proposing. While the subliminal suggestion is to defray action on all forms of so-called modernist high culture, are we not led into this forlorn conclusion precisely by dint of the wonderful scientific work conducted by Mayr and his colleagues at Harvard, which has an annual budget of several billion dollars? So, is it not imperative that solutions to the dangers posed by the culture of enlightenment are to be found within enlightenment itself? Obviously, there cannot be an immediately satisfying answer to this question either way. So, let me ask a series of rhetorical questions to conclude the discussion.

Can we not view the otherwise wonderful results from Harvard as a *reductio* to the effect that this knowledge need not be pursued anymore? Elizabeth Kolbert has remarked with some irony that let us not ask the scientific question of when the human species might become extinct, because we might be extinct before we reach a definite scientific answer (Drake 2015). Sensible people have started advocating the disarming of the planet. Does that not amount to the demand that the knowledge-systems that go into the construction of weaponry—from pistols to hydrogen bombs—be deliberately set aside? I am told that the Japanese monarchs refused to introduce guns in their army for centuries even though the Europeans have been trying hard to sell them the lucrative technology. The reason was, in a battle with swords, you have to face another human being from close quarters; so you are compelled to confront the moral issue of killing a human being. In a gun battle from a distance, you do not face that moral choice.

Why should that argument not extend to the knowledge of making cars and aeroplanes, since these technologies require the extraction of bauxite from revered mountains? Once we get the feel of the mess into which modern living has pushed the planet, why should we stop at cars and aeroplanes? Why not computers, mobile phones, skyscrapers, libraries, orchestras, art museums, cities and asphalt roads? The children of the gods of Niyamgiri lived without them happily for thousands of years. Exactly what argument do we have for not emulating their lives in full?

References

Bera, S. 2013. Niyamgiri answers. *Down to Earth*, 31 August, viewed on 1 May 2016, http://www.downtoearth.org.in/coverage/niyamgiri-answers-41914.

Chomsky, N. 2003. *Hegemony or Survival*. New York: Metropolitan Books.

Chomsky, N. 2005. Manipulation of fear. Foreword Essay, In N. Mukherji *December 13: Terror Over Democracy*. New Delhi: Bibliophile South-Asia.

Crow, T.J. 2010. The nuclear symptoms of schizophrenia reveal the four quadrant structure of language and its deictic frame. *Journal of Neurolinguistics* 23(1): 1–9.

de Queiroz, K. 2005. Ernst Mayr and the modern concept of species. In *Proceedings of the National Academy of Sciences*, 201, 6600–07, viewed on 1 May 2016, http://www.pnas.org/content/102/suppl_1/6600.full.

Drake, N. 2015. Will humans survive the sixth great extinction? *National Geographic*, 23 June, viewed on 1 May 2016, http://news.nationalgeographic.com/2015/06/150623-sixthextinction-kolbert-animals-conservation-science-world/.

Foreman, D. 2004. *Rewilding North America: A Vision for Conservation in the 21st Century.* London: Island Press.

Freire, P. 2005. *Pedagogy of the Oppressed.* New York: Continuum.

Gettys, T. 2014. Noam Chomsky on human extinction: The corporate elite are actively courting disaster. *Raw Story*, 18 June, viewed on 1 May 2016, http://www.rawstory.com/20-14/06/noam-chomsky-on-human-extinctionthe-corporate-elite-are-actively-courting-disaster/.

Kolbert, E. 2014. *The Sixth Extinction.* New York: Picador.

Kothari, A. 2015. Revisiting the legend of Niyamgiri. *The Hindu*, 2 January.

Lukacs, M. 2013. Noam Chomsky slams Canada's shale gas energy plans. *Guardian*, 1 November, http://www.theguardian.com/environment/2013/nov/01/noam-chomsky-canadas-shale-gas-energy-tar-sands.

Mayr, E. 2001. *What Evolution Is.* New York: Basic Books.

Mukherjee, H.B. 2013. *Education for Fullness: A Study of the Educational Thought and Experiment of Rabindranath Tagore.* New Delhi: Routledge.

Mukherji, N. 2012. *The Maoists in India: Tribals Under Siege.* London: Pluto Press.

Mukherji, N. 2016. Education for the Species. *Economic and Political Weekly*, 51(32), 06 Aug.

Soulè, M.E. 1996. What do we really know about extinction? In *Genetics and Conservation*, C.M. Schonewald-Cox, S. Chambers, B. MacBryde, and L. Thomas (Eds.), 111–143. Menlo Park, CA: Benjamin-Cummings.

Striedter, G.F. 2004. *Principles of Brain Evolution.* Sunderland, MA: Sinauer Associates Inc.

Tattersall, I. 2012. *Masters of the Planet: The Search for Our Human Origins.* New York: Palgrave Macmillan.

Vanaja, S. 2014. Lingraj Azad, warrior of Niyamgiri. *Round Table India*, 21 April, viewed on 1 May 2016, http://roundtableindia.co.in/index.php?option=com_content&view=article.&id=7375:lingraj-azad-warrior-of-niyamgiri&catid=119:feature&Itemid=132.

Wood, G. 2015. What ISIS really wants. *Atlantic*, March, viewed on 1 May 2016, http://www.theatlantic.com/magazine/archive/2015-/03/what-isis-really-wants/384980/?utm_source=SFFB

Index

Printed in the United States
By Bookmasters